Ethics, Qualitative And Quantitative Methods In Public Health Research

Mbuso Mabuza

Published by Mbuso Mabuza, 2022.

ETHICS, QUALITATIVE AND QUANTITATIVE METHODS IN PUBLIC HEALTH RESEARCH

First edition. December 1, 2022.

ISBN: 979-8215492062

Written by Mbuso Mabuza.

Also by Mbuso Mabuza

A Healthy Mind And Best You: Achieving Great Results in Every Aspect of Your Life
Purposeful And Better You
Sustainable Development Calls for Effective Strategic Leadership for Efficient Health Systems
Health Promotion In Low Socioeconomic Settings
Medicine and Sociology of Health
Qualitative Methods In Public Health Research
Epidemiological Research
Ethics, Qualitative And Quantitative Methods In Public Health Research
Global Health Disaster Management
Global Health Policy And Programme Challenges
The Journey of Life Has a Gift of Purpose

Table of Contents

Preface 1

PART 1: ETHICS AND QUALITATIVE METHODS IN PUBLIC HEALTH RESEARCH 3

Chapter 1 4

Chapter 2 14

Chapter 3 36

Chapter 4 64

Chapter 5 74

Chapter 6 110

PART 2: QUANTITATIVE METHODS IN PUBLIC HEALTH RESEARCH 123

Chapter 7 124

Chapter 8 139

Chapter 9 163

Chapter 10 181

Chapter 11 192

Chapter 12 203

To my beloved mother and father

Preface

This book provides an introduction to ethics, research design as the most important part of the qualitative research process, the importance of theoretical frameworks and the relationship between the researcher and the researched in the qualitative research process.

It also provides an introduction to epidemiological research and statistical methods covering key concepts and all the main types of epidemiological study. It elicits a critical understanding of the purpose and context of quantitative research including the basis for selecting appropriate research designs from a thorough grounding in the uses and methods of epidemiology; key statistical concepts and techniques needed for the basic analysis of data; critical evaluation of statistical and epidemiological techniques in health research.

The book has been designed around a number of core published information dealing with various topics from the United Kingdom, Europe and the low- and middle-income countries. These studies have been chosen to introduce you to a wide range of study methods. The main intriguing aspects of their design provide examples which are used to help you understand the fundamental principles of good research, and to practise these techniques yourself.

The book is divided into two sections. Section 1 deals with ethics and qualitative methods in public health research. This section is organised according to the following chapters: ethics and introduction to qualitative research; qualitative research study design; qualitative methods – participant observation; qualitative methods – interviews, texts, and qualitative diaries; participatory and action research; data analysis – coding, storing and managing qualitative data; analysis –

interpretation; validity and presentation of findings to different audiences.

Section 2 of this book deals with quantitative methods in public health research. This section is divided into the following chapters: scientific method and introductory concepts; routine data sources and descriptive epidemiology; surveys; cohort studies; case-control studies; and critical appraisal of research evidence.

This book will be a valuable resource to health professionals, researchers, statisticians, data scientists, health programmers, policymakers, medical students, graduate and postgraduate students in public health and related disciplines.

Dr Mbuso Mabuza

Global Health, Medicine and International Public Health Consultant

PART 1: ETHICS AND QUALITATIVE METHODS IN PUBLIC HEALTH RESEARCH

Chapter 1

Introduction to Ethics and Qualitative Methods in Public Health Research

There are various definitions of ethics, but all these definitions generally point towards a moral discourse which is sometime intertwined with a regulatory discourse (Ashcroft, 2003). When researchers conduct research, whether quantitative or qualitative in approach, the moral discourse should take precedence and ethical principles should be integral especially when the research deals with human subjects.

Different writers express ethical principles differently, but they are all important in carrying out qualitative research. For example, the Belmont Report of 1979 expounds the four ethical principles of autonomy, beneficence, non-maleficence, and justice (van Niekerk & Kopelman, 2005). Whereas, other writers expound just three ethical principles such as respect for persons, which includes and considers the principle of autonomy, while the second principle of beneficence considers and includes the principle of non-maleficence, and lastly the third principle of justice (Joubert & Ehrlich, 2007).

For example, there are a number of ethical issues around HIV testing for pregnant women. The principle of autonomy may be problematic and very possibly violated here, because the women may be tested for HIV because they do not have a choice, and this issue needs more attention (King et al, 2013). One has to bear in mind that autonomy from a western world construct is different from the developing world where personal choice is extremely limited (van Niekerk & Kopelman, 2005). It is hardly surprising that some researchers argue that the issue of autonomy in HIV testing for pregnant women in developing countries needs to be taken seriously, as there is serious doubt that the consent obtained from them is genuine and that there is an element of coercion in the antenatal counselling services (van Niekerk & Kopelman, 2005). The subjects such as the pregnant women may be so vulnerable such that they may fear that their healthcare would be affected if they decline to participate in the research (World Health Organisation, 2003).

When international organisations conduct research such as HIV and AIDS research in low- and middle-income countries such as in sub-Saharan Africa, the subjects may have expectations that they would benefit something out of it, only to find that there is nothing really to benefit from such research, instead it is that particular international organization that benefits by obtaining so much information about the subject matter, and this may cause mental harm to the subjects (van Niekerk & Kopelman, 2005). This shows that the principles of maleficence and beneficence are violated in this regard.

It seems as if pregnant women in low- and middle-income countries such as southern Africa, are chosen for HIV testing because they are an easy target from which the national HIV prevalence can be estimated. This is unfair, and it shows that the ethical principle of justice is also violated.

The main principles for carrying out qualitative research include the principles of autonomy, non-maleficence, beneficence, and justice now threaten to undermine the research enterprise (Pollock, 2012). The issue of HIV testing for women in low- and middle-income countries such as in southern Africa, seems to be a violation of all the ethical principles mentioned above, and therefore such research needs a distinctive approach in such settings (Ahn et al, 2003).

Given the ethical dilemma surrounding research conducted on humans in low- and middle-income countries, one is left to wonder where the core of the problem really lies and how it may be addressed. Does the problem lie with the differences in autonomy between different countries or with the lack of values among those who conduct and sponsor research? It is not that easy to put the finger on the pulse of the problem because the real cause of the problem appears to be multi-faceted, but its nucleus appears to be embedded in conflict of interest between research breakthrough and the public good. It is sometimes acknowledged that the biggest challenge is that a large part of ethics is undetectable (Joungtrakul & Allen, 2012).

It cannot be denied that research has solved a number of health issues in many parts of the world, and the emergence of new diseases and the proliferation of epidemics such as HIV and AIDS sometimes require that further research be done in order to get a better understanding for better solutions. But for further research to happen, funding is required, but the unfortunate situation is that most of those who conduct research and those who sponsor it claim to uphold the highest ethical standards, in theory or on paper, but practically, this is often not the case, as the researchers may find themselves in a tight spot due to conflict of interest (Tilley & Woodthorpe, 2011).

Sometimes, unethical research is disguised in the name of social responsibility, that even seasoned researchers may be pressurized by the sponsors' demands to consider other ways of research which fall short of meeting the acceptable ethical standards (Kotler & Lee, 2005). It is widely recognized that some sponsors of research may have a vested interest in the results of the research, and as a result they may sway the way the research is conducted to suite their hidden objectives, and in that process inadvertently compromise the highest ethical standards (Alsmadi, 2010). Even if the set ethical regulations or ethical standards are adhered to, the concern still remains because most of the time the ethical standards are based on a western model which may not be fully applicable to a developing world's situation.

This calls for serious reflection, as some scholars question whether the proliferation of health research in developing countries is conducted with sincerity or with hypocrisy (Kilian & Hennings, 2011). Perhaps, the problem of autonomy in different contexts and settings could be addressed by ensuring that international research ethics are rationally constructed as a negotiated moral order that respects culturally and individually accepted aspects of non-negotiability (Baker, 1998). Community-specific ethical standards should be developed in collaboration with the respective communities. There must be guarantee that the research will not cause any harm to the participants or community, and that results are only published after review and approval by the community (van Niekerk & Kopelman, 2005).

It is so unfortunate that the principle of autonomy or respect for persons seems to be pronounced in theory but not often applied in practice especially when research is conducted in low- and middle-income countries. This is surprising, because, one would think that the principle of autonomy which aims to protect those with diminished autonomy and expounds the interests of the individual over the interests of science or research should be held in high regard (Joubert & Ehrlich, 2007).

Interestingly, some researchers report that western societies are generally powerful, privileged and highly legalized (van Niekerk & Kopelman, 2005), and the citizens of such countries are generally free to make their own decisions at individual level (Olivier & Fishwick, 2003). On the other hand, in low- and middle-income countries such as in sub-Saharan Africa, autonomy differs because it usually depends on what the community generally deems as acceptable or not acceptable or based on the fact that people exist only in relation to others, such as families, tribes, and constituencies especially in the rural areas (van Niekerk & Kopelman, 2005). For example, the fact that sub-Saharan Africa is regarded as the epicentre of the HIV and AIDS epidemic, subjects in the sub-Saharan countries may easily be coerced to participate in HIV research (Ahn et al, 2003). A lot of people volunteered to participate in HIV research including the medical male circumcision research and the antimicrobial vaginal gel research in South Africa, perhaps, because of the perceived risk benefit by the community. This leaves us with doubts about the autonomy of subjects from low- and middle-income countries such as South Africa.

The issue is that regulations are very stringent about the adherence to ethical principles when conducting research in the western world, whereas, the unfortunate or desperate situation that developing countries find themselves in, makes the application of stringent ethical regulations or principles very questionable in the developing world (van Niekerk, 2005). It is a travesty that the fragile health systems, burden of disease and poverty in the developing countries is often used as justification to conduct research in these countries, whereas, such studies could not pass even the slightest ethical rigour in the sponsoring western country (van Niekerk & Kopelman, 2005).

This suggests that the western world's construct of autonomy is generally individualistic and legalistic, unlike the African construct of autonomy which is generally relational and extended. It is interesting that other researchers also add that subjects in the developing world sometimes feel pressured, powerless and that they may also volunteer to participate in research because they may feel a sense of duty to the researcher or health professional (Richards & Shwartz, 2002).

The western world's construct of autonomy is powerful at individual level unlike the construct of autonomy in developing countries which is relational and with extremely limited personal choice (King et al,

2013). This raises serious ethical issues because research on human subjects in low- and middle-income countries could be open to exploitation by the sponsoring western country.

Upholding the utmost ethical standards and principles could be a challenge when conducting research involving human subjects, and the challenge could be more pronounced when conducting research on vulnerable groups or special populations. According to Cromar et al (2006), vulnerability has many definitions, but, perhaps, the simplest is the capacity for loss, and this may include groups or populations with low levels of personal choice such as people of low socioeconomic status, people with disabilities, children, the elderly, and minority groups, among others.

This suggests that if we subscribe to the highest ethical standards and principles, research on vulnerable groups or special populations should best be avoided. However, it is possible that that there may be instances where research on vulnerable groups or special populations could be inevitable. In such instances, the approach of the researcher should be of the utmost ethical standards.

The researcher should first get a better understanding of the circumstances of the vulnerable groups or special populations in their own setting, including their culture and the way they view their life (Heath et al, 2007). It is also imperative that the researcher fully explains truthfully what the proposed research is all about, in a manner that will be easier to understand by the targeted vulnerable groups. Emphasis is placed on the central importance of obtaining consent and to maintain absolute confidentiality (Heath et al, 2007) and to share research information with the target population without sensationalizing, to ensure that the research does not cause any harm to the subjects (Harmon & Chen, 2012).

In the case of conducting research on children, consent should be obtained from their parents or guardians, but the legal age of consent may vary from country to country as some countries permit individual consent for children who are at least 14 years of age while some require the child to be at least 18 years of age (Ensign, 2003). People with disabilities may give their consent as individuals as long as they are at the legal age of consent, but for those who are mentally affected, the researcher may be required to obtain consent from the relevant institution or organization to which the affected subjects are affiliated, as laid down by the law (Touitou et al, 2004). The researcher should always strive towards ensuring that the subjects are not coerced to participate in the research. This point is alluded to by some researchers who emphasize that that voluntarism is a significant precondition for obtaining truly informed consent (van Niekerk & Kopelman, 2005).

It is an imperative for the researcher to ensure that he or she establishes healthy research rules and boundaries without compromising the interests of the vulnerable groups. Whilst there is often the urge for researchers to make a scientific breakthrough, the danger is that this may end up overstretching the boundaries of ethical standards and principles, at the expense of the vulnerable groups or special populations.

References

Ahn, M.J., Grimwood, A., Schwarzwald, H. & Herman, A. (2003) 'Ethics and the AIDS pandemic in the developing world', *Journal of International Association of Physicians in AIDS Care*, 2 (2), pp. 81-87.

Alsmadi, S. (2010) 'Marketing research and social responsibility: ethical obligations toward the society', *Journal of Accounting – Business & Management*, 17 (1), pp. 42-47.

Ashcroft, R.E. (2003) 'The ethics and governance of medical research: what does regulation have to do with morality?', *New Review of Bioethics*, 1 (1), pp. 41-58.

Baker, R. (1998) 'A theory of international bioethics: the negotiable and the non-negotiable', *Kennedy Institute of Ethics Journal*, 8 (3), pp. 233-273.

Cromar, N., Cameron, S. & Fallowfield, H. (2006) 'Environmental epidemiology'. In: *Environmental health in Australia and New Zealand*. Melbourne: Oxford University Press

Ensign, J. (2003) 'Ethical issues in qualitative health research with homeless youths', *Journal of Advanced Nursing*', 43 (1), pp. 43-50.

Harmon, S.H.E. & Chen, K. (2012) 'Medical research data-sharing: the public good and vulnerable groups', *Medical Law Review*, 20 (4), pp. 516-539.

Heath, S., Charles, V., Crow, G. & Wiles, R. (2007) 'Informed consent, gatekeepers and go-betweens: negotiating consent in child- and youth-orientated institutions', *British Educational Research Journal*, 33 (3), pp. 403-417.

Joubert, G. & Ehrlich, R. (2007) *Epidemiology: a research manual for South Africa.* 2nd ed. Cape Town: Oxford University Press Southern Africa.

Joungtrakul, J. & Allen, B.M. (2012) 'Research ethics: a comparative study of qualitative doctoral dissertations submitted to universities in Thailand and the USA', *Science Journal of Business Management*, 2012 (2), pp. 1-11.

Kilian, T. & Hennings, N. (2011) 'Corporate social responsibility between hypocrisy and sincerity: a longitudinal analysis in annual reports from 1998-2009', *American Marketing Association*, Summer 2011.

King, E., Maman, S., Wyckoff, S., Pierce, M. & Groves, A. (2013) 'HIV testing for pregnant women: a rights-based analysis of national policies', *Global Public Health*, 8 (3), pp. 326-341.

Olivier, S. & Fishwick, L. (2003) 'Qualitative research in sports sciences: is the biomedical ethics model applicable?', *Forum Qualitative Social Research*, 4 (1), pp. 1-15.

Pollock, K. (2012) 'Procedure versus process: ethical paradigms and the conduct of qualitative research', *BMC Medical Ethics*, 13 (25), pp. 2-12.

Richards, H.M. & Schwartz, L.J. (2002) 'Ethics of qualitative research: are there special issues for health services research?', *Family Practice*, 19 (2), pp. 135-139.

Thompson, A.K., Faith, K., Gibson, J.L. & Upshur, R.E.G. (2006) 'Pandemic influenza preparedness: an ethical framework to guide decision-making', *BMC Medical Ethics*, 7 (12), pp. 1-11.

Tilley, L. & Woodthorpe, K. (2011) 'Is it the end for anonymity as we know it? A critical examination of the ethical principle of anonymity in context of 21st century demands on the qualitative researcher', *Qualitative Research*, 11 (2), pp. 197-212.

Touitou, Y., Portaluppi, F., Smolensky, M.H. & Rensing, L. (2004) 'Ethical principles and standards for the conduct of human and animal biological rhythm research', *Chronobiology*, 21 (1), pp. 161-170.

Van Niekerk, A. A. & Kopelman, L.M. (2005) *Ethics & AIDS in Africa: the challenge to our thinking*. Claremont: davidphilip

World Health Organisation (2003) *Principles and practices: the implementation of ethical guidelines for research on HIV*. World Health Organization Department of HIV/AIDS, 2-3 June 2003. Geneva.

Chapter 2

Qualitative Research Study Design and Methods

The main focus of this chapter is on the following topics:

- Introduction to qualitative research study design and methods
- Ethnography
- Theoretical and purposive sampling
- Covert methods and ethical issues
- Field notes

2.1 Introduction to Qualitative Research Study Design and Methods

POSITIVISM BEING A model of modern scientific enquiry puts emphasis on elements such as empiricism, unity of method and value-free enquiry (Green & Thorogood, 2009). However, there is growing criticism of the positivism approach as it is being regarded as an unsuitable theory of knowledge for qualitative researchers (Darbyshire et al, 2005).

Empiricism as one of the elements of positivism suggests that only observable phenomena or what can be measurable or broken down into measurable components should be studied (Green & Thorogood, 2009). This may be a problem in qualitative research because not everything can be broken down into quantifiable or measurable components. For example, patients who receive treatment from a complementary health professional such as a Chiropractor or a

Naturopath or a Homoeopath, may feel much better, but the mechanism by which the therapy makes them feel better may not be measurable by quantitative research methods. Whereas, qualitative researchers may be interested in understanding how the patients feel after receiving the complementary therapy.

Unity of method as the second element of positivism suggests uniform standards, laws or methods for scientific enquiry (Green &Thorogood, 2009). This is unsuitable in qualitative research because interpretativists acknowledge the uniqueness of different individuals or situations, which implies that no one glove fits all. For example, while the positivism theory would look at cause effect such as the amount of isocyanate exposure that would cause asthma, qualitative researchers would like to look at the quality of life of the people who are exposed to isocyanates and how they perceive themselves individually or as a group in relation to their environment.

Value-free enquiry is the third element of positivism which suggests that the researcher is divorced from the research and is as objective and as neutral as possible that she does not influence the outcome of the research (Green & Thorogood, 2009). Qualitative researchers such as interpretativists acknowledge that the researcher cannot be completely objective in the research especially when studying human beings, because human beings are complex and unpredictable. The outcome of the research could be influenced by who does the research, how the research is done, and how the researcher interacts with or connects with the research participants.

The interpretative and social constructionist theories of knowledge identify with qualitative research (Green & Thorogood, 2009). The interpretative theory has the premise that human beings are unique and they differ in the way they interpret the reality of the world (Green & Thorogood, 2009). The social constructivist theory acknowledges that people's experiences are constructed through historical and socio-political processes, which implies that social constructivist researchers cannot ignore people's background, values, traditions, and the way they are governed or led (Alderson, 1998). Although there seems to be an overlap between the interpretative and social constructionist approaches, individuals' perception of how they feel or think is more important than putting them in one basket.

The Marxist theory focuses on the researcher as the voice of the researched and the need for their empowerment (Lather, 1986). Much as it is important to empower the researched, there is an issue with the Marxist approach because of its categorization of people according to class and economic structures, which undermines the uniqueness of individuals, regardless of their socio-economic circumstances (Becker, 1984). Positivism is unsuitable for qualitative research.

According to Larkin (2011), both Marxism and social constructivism are sociological concepts. However, there are differences between Marxism and social constructivism.

Marxism or the materialistic approach is critical of capitalism particularly with regard to the exploitation and domination of people that are considered to be of low class by those that are considered to be of high class and who have power and wealth (O'Mahony, 2011). This suggests an inevitable conflict produced between those who own the means of generating wealth and those who have to sell their labour (Green & Thorogood). Questions that may be asked include the relationship between poverty and ill-health, low social class and ill-health, and the effects of globalization (Green & Thorogood, 2009). A typical concern in post-apartheid South Africa is the deterioration of socio-economic conditions of the majority who are considered to be in the lower class which could be accompanied by ill-health, whereas the super-rich minority may enjoy good health as they can afford better healthcare.

When it comes to social constructivism, how we divide up the world in order to comprehend it is the result of historical, social and political processes rather than an inevitable result of our greater understanding of the reality of the body, or disease (Green & Thorogood, 2009). For example, in post-apartheid South Africa, many women and people of African origin may not realize that they have so many possibilities to realize their full potential because they had already learned to categorise themselves as disadvantaged due to the legacy of apartheid. Kitchin (1998) adds that people with disabilities were not born with a predisposition to recognize and categorise themselves or others, but they learn to categorise themselves as oppressed through historical and socio-politico-cultural processes which is not necessarily related to capital concerns. Kitchin (1998) is interesting with the arguments that disability is spatially socially constructed, and the observation that disabilities are spatially-socially constructed and that social relationships currently spatially isolate and marginalize people with disabilities.

The Marxism and social constructivism theories may appear as if they share something in common, but they are actually different. Marxism is critical of capitalism and the ideals that grounded the Marxist philosophy are political action and social responsibility (Jefson, 2005). Social constructivism assumes that reality is socially constructed (Green & Thorogood).

There is an upsurge of studies that employ a combination of the different epistemological dimensions such as quantitative and qualitative approaches (Moffat, 2006). The upsurge in the mixed or multi-method approach is probably due to the growing realization even from proponents of the positivist approach that most public health problems are deeply embedded in a social, political and economic context (Moffat, 2006). This point of view is further confirmed by the fact that a multi-method approach is becoming more common as it is encouraged by funding bodies, and is often used in pilot studies, or to add depth or power to quantitative studies, or for the purposes of addressing different aspects of the same research question (Green & Thorogood, 2009).

For example, a recent study on family planning methods in rural, urban and industrial settings of southern Africa, included a structured survey of knowledge attitudes and practices (quantitative approach), in addition to in-depth interviews (qualitative approach). This multi-method approach was valuable because it gave a broader perspective that would not have been possible if only one approach was used.

Some researchers observe that since there is not one research paradigm that can tap all dimensions of complex research problems, a mixed approach could be of some benefit because this could allow the drawing of conclusions from a synthesis of results (Ulin, 2005). It is further added that mixed or multiple approaches that involve a positivist or quantitative approach and a qualitative approach could create a broader perspective on the research problem and often more persuasive findings for policymakers (Ulin, 2005).

On the other hand, there could be challenges which may arise if the quantitative or positivist approach yields completely different results from the qualitative approach as this could compromise the credibility of the results (Moffat et al, 2006). Such discrepancy could also be an advantage because it allows for an opportunity to interrogate the different results more fully (Moffat et al, 2006).

There are some benefits of a mixed approach, but there are also challenges as philosophies or theories that define positivism are incompatible with the philosophies that define qualitative research. It is therefore advisable that even if the research is conducted from a multi-method approach that involves both a qualitative perspective and positivist or quantitative perspective, the analysis of the data is done separately according to the principles of each epistemological dimension (Maree, 2010).

Positivism can have a role in qualitative inquiry if it is combined or merged with other theories suited for qualitative inquiry such as the social constructionist theory, but this merging of different epistemological dimensions is not without challenges.

2.2 Ethnography

IT HAS TO BE EMPHASIZED that ethnography is firmly rooted in the inductive approach whose purpose is to interpret the social world the research subjects inhabit in the way in which they interpret it (Saunders et al, 2003). Green & Thorogood (2009) observe that the main disadvantage with being purely a detached observer is that the setting is initially strange. In this regard, if the ethnographer remains detached from the events happening to keep his/her objectivity and not get personally involved in the community he/she is researching, he/she deprives himself/herself as there is a wealth of rich information that he/she could have obtained to help the course of his/her research.

In ethnography, the researcher should stop observing while sitting on the tree but should come down and immerse himself/herself in the culture of the community he/she researches. This immersion will enable the researcher to have a feel of what the research subjects experience and this gives the researcher the flexibility and responsiveness to change since the researcher will constantly be developing new patterns of thought about what is being observed

(Saunders et al, 2003). The concept of ethnography or participant observation signifies the relation which the human observer of human beings cannot escape having to penetrate in some fashion in the experience and actions of those he/she observes (Liampuntong & Ezzy, 2005).

Lerum (2001) observes that the problems of objectivity in ethnographic research are rooted in the larger issue of emotional detachment is implicit in the standard scientific method, which is facilitated by academic armor, and the problem at hand is that researchers routinely closet their preferences behind a wall of objectivity. Dropping the academic armor enables researchers to emotionally engage with the research participants, and knowing research participants intimately is transformative for researchers and hence, apparently for their conclusions (Lerum, 2001). It would greatly benefit researchers to drop the academic armour and immerse themselves in the richness of interconnectedness with the community they are researching. It is highlighted that the interconnectedness between self and society illustrates the impossibility of any researcher being completely objective (Lerum, 2001).

The notion that "an ethnographer should remain purely a researcher from the start to the end of an ethnographic study, 'detached' from the events happening to keep his/her objectivity, and should not get personally involved in the community he/she is researching."

"I joined them in their setting, I walked with them, I ate with them, I learned their language, and I learned their culture because I wanted to have a deeper understanding of their way of life from their world view."

The foregoing is typically what an ethnographer does in participant observation. According to Welman et al (2005), ethnography is a descriptive design that aims to investigate amongst individuals, social groups, or community, and its focus is on the behavioural regularities of everyday situations.

The fact that the researcher spends time in the field and becomes part of the social group or community for which the research is focused is in a better position to gain a deeper understanding of the social group or community (Liampuntong & Ezzy, 2005). This deeper understanding and the fact that the researcher sees the research from the people's own perspective could minimize the likelihood of misinterpretation of the social group or community studied (Liampuntong & Ezzy, 2005). Gaining a depth of rich information from the people's own world view could enhance the formulation of questions that would make sense to the group or community studied as these questions will be in the people's native language (Liampuntong & Ezzy, 2005). Ethnography could also be an advantage in the sense that the deeper understanding of the group or community could be used to enhance interventions aimed at bringing about change (Liampunting & Ezzy, 2005). For example, it may be used to bring about behaviour change to reduce the spread of HIV in many parts of sub-Saharan Africa where the burden of HIV and AIDS is already so huge.

Disadvantages may include the challenge of gaining access to the community or group, due to multiple gate keepers (Green & Thorogood, 2009). Traditional leaders are influential gate keepers in many parts of sub-Saharan Africa. Another disadvantage is possible premature termination of the research due to the fact that the researcher may have discomfort about the conflict between what she sees from her own world view and what the people or informants see from their own world view (Liampuntong & Ezzy, 2005). It may also be a disadvantage if the researcher is required to spend a lengthy period in the field especially if there are budget constraints, and sometimes the researcher may be physically or emotionally affected (Liampuntong & Ezzy, 2005). Although the concepts of the research findings may be used to inform the researcher's understanding of similar contexts, it may be a disadvantage that the information may not be generalizable or transferrable to other settings (Green & Thorogood, 2009).

Ethical implications of carrying out ethnographies may include the issue of emotional harm to the community or group studied especially if the research results are viewed as inaccurate or non-beneficial or malicious or embarrassing (Ulin et al, 2005). Green & Thorogood (2009) highlight that it could be unethical if influential gate keepers permit the researcher to enter the community and conduct research without the individual informed consent of the participants. As a result, there could be emotional harm to all or some of the community or group members.

According to Liampuntong & Ezzy (2005), the most difficult part of ethnography is gaining entry into a site or community. Green & Thorogood (2009) mention multiple gate keepers as a challenge, and define formal gatekeepers as those people whose permission is required before fieldwork can commence, and informal gatekeepers as those people whose support is crucial for fieldwork to be practically possible.

Identifying formal gatekeepers in institutions may be easy, but gaining access may not be straight forward as the researcher may need to have internal networks to smooth the way (Green & Thorogood, 2009). For example, writing a letter to the Chief Executive of a corporate organisation may seem appropriate to request permission to conduct research within the organisation, but there is a possibility that such a letter may be ignored or overlooked by the intended recipient who often has other pressing business matters to attend to. Perhaps, it may be easier to get permission to do research in the organization if the researcher is introduced to the Chief Executive by an internal person such as the Corporate Affairs Manager who is responsible for the public relations aspect of the business.

However, if the Corporate Affairs Manager is not convinced that the intended research will benefit the organization, it is very unlikely that the request will even be taken to the Chief Executive for his approval, bearing in mind that the Corporate Affairs Manager looks after the good image of the organization. In such a workplace setting, informal gate keepers would very likely be the trade union representatives and the culture charter champions, without whose buy-in, the research would not be practically possible. This shows that the researcher will have to go through multiple gate keepers in large institutions. However, it may be easier to gain access to small institutions as these institutions may see it as an honour to be chosen for the research (Green & Thorogood, 2009).

It may not be easy to identify formal gatekeepers in the community especially if the researcher is not familiar with such communities (Green & Thorogood, 2009). It would therefore, be important for the researcher to do her homework thoroughly so that she could be in a position to know whom to contact for permission. For example, it would be a good idea to first get the buy-in of traditional leaders of the rural parts of Southern Africa if the research is intended for those communities (van Niekerk & Kopelman, 2005). Informal gate keepers in the community could possibly be the community development volunteers and team leaders.

Access to a site or community is an issue that should be seriously taken into account as it could determine the fate of the ethnographic study. Ulin et al (2005), highlight the need to recognize the variations of the challenge to access different sites.

2.3 Theoretical and purposive sampling

THERE ARE VARIOUS SCHOOLS of thought about qualitative sampling, but the different variants, the overlaps and the synonymous use of theoretical sampling and purposive sampling by some authors could create confusion for the novice researcher (Coyne, 1997).

Theoretical sampling is part of the grounded theory which is one of the influential methodologies used in qualitative research (Liampuntong & Ezzy, 2005). Theoretical sampling follows a gradual approach as theory builds until the point of saturation because the researcher may not know in advance what to sample and where this will lead her (Coyne, 1997). The origins of theoretical sampling seem to have been at the time of the discovery of the grounded theory by Glazer and Strauss in 1967, and during that time, the prevailing paradigm was quantitative research even in the field of Sociology (Coyne, 1997). Given the prevailing paradigm at the time, the pioneers of theoretical sampling and of the grounded theory in general were under pressure to conform by rigorously verifying their new research approach through quantitative means (Coyne, 1997).

As part of the grounded theory, theoretical sampling identifies with the three pillars of joint collecting, joint coding and joint analysis of data as new theory emerges until all the codes and categories have been saturated. While there are different variants of qualitative sampling, purposeful sampling is often regarded as part and parcel of theoretical sampling (Coyne, 1997). This is so because in theoretical sampling, the researcher initially zooms in and indiscriminately selects an area or a sample where a certain phenomenon exists, and as data analysis proceeds and as new theory emerges, the researcher will purposively select those whom she considers will provide data of relevance to the emerging theory until a point of saturation is reached (Coyne, 1997).

Purposive or theoretical sampling is important for qualitative research because it creates an opportunity for obtaining rich sources of data and the flexibility and open-endedness of the interviews encourages the emergence of new conceptual ideas (Green & Thorogood, 2009). Through purposive or theoretical sampling, the creativity and adaptability of the qualitative researcher enables a view of the real life situation so that it makes sense to the researched, and also attempts to be applicable to other contexts rather than the fixed perspective of quantitative research (Coyne, 1997). Other authors take this further and highlight that qualitative research emphasizes insight rather than generalizability, and depth rather than breadth, and through purposive or theoretical sampling, this aim is achieved (Ulin et al, 2005). We do not pre-determine the sample before we begin our fieldwork because if we did, we could be confined to the same type of knowledge as we would not be able to take the emergence of new theories further.

2.4 Covert methods and ethical issues

MUCH AS ONE HAS TO acknowledge the ethical issues related to covert status, there is a role for covert methods in observational studies. Covert observation may assume a variety of guises, and it basically amounts to the researcher deceiving the participant about the actual purpose of the research or about the identity of the researcher (Mouton, 2002). Covert research is particularly applicable in studies where participant observation or in-depth interviewing is used (Mouton, 2002).

It has to be acknowledged that in participant observation studies covert research is often regarded as intrinsically unethical in comparison with overt research, largely due to its deceptiveness and non-disclosure to subjects (Spicker, 2011). Green & Thorogood (2009) highlight that covert research clearly raises ethical dilemmas because the autonomy of the participant is not respected and obtaining informed consent is virtually impossible prior to the commencement of the fieldwork. For example, I remember while we were still growing up in Swaziland, we used to see some people from overseas who had come to preach and to live among the community and they even adopted Swazi names. After approximately two years when they were about to leave, it transpired that they had actually come to do their research in Swaziland. One can now realize that there was an ethical issue of non-disclosure as the research was done under cover, and informed consent was not obtained.

However, there is an interesting argument by some authors who observe that the moral high ground that is often attributed to overt research is questionable as most overt studies employ covert practices (McKenzie, 2009). It is also argued that covert research has an important role to play in observational studies as it can be a useful strategy to counter the guinea-pig effect (Mouton, 2002). Herrera (1999) takes this point further with the argument that the omission of an informed consent form in covert methods is on the basis of protecting the participants from apprehension, nervousness or even criminal prosecution. Saunders et al (2003) see covert methods as a solution to the issue of reactivity in overt observation studies. Other authors also argue that much as the condemnation of covert methods is a legitimate concern in terms of the rights of subjects, such condemnation is over-generalised because research subjects are not the only people with rights (Spicker, 2011). Moreover, it is also argued that much as the subjects are entitled to their privacy, it is widely accepted that after the covert observation has taken place those affected should be informed about what has happened and why, and this process is known as debriefing (Saunders et al, 2002).

Much as covert research raises many ethical dilemmas, there is counter-argument that covert research can still be applicable in observational studies, where it is practical. The dilemma confronting the researcher is how to weigh the moral interests of the subject against the interest of research or science (Mouton, 2002).

2.5 Field notes

PLACE: TOWN X MINI-bus station

Observer: YYZ Mabuz

Date: 18 May 2022

Time: 6.15am to 6.45am

Observation	Time
I am at Town X mini-bus taxi rank in the KwaZulu-Natal province of South Africa, and have just boarded a mini-bus taxi. I am the first passenger to board the mini-bus taxi and am sitting at the back seat while I watch other passengers boarding the same mini-bus taxi destined for a local town called Town Y.	6.15am
Two African young men probably in their mid-twenties were standing next to the entrance of the mini-bus taxi, seemingly surprised and unimpressed that the mini-bus taxi was virtually still empty. They complained between themselves that they were going to be late for work	6.17am
Three other African passengers (a woman probably in her late sixties, a man probably in his early seventies and using a walking stick, and a young woman probably in her mid-thirties) arrived and they enquired in the Zulu language from the two young men standing next to the mini-bus taxi about whether this was the taxi destined for Town Y. The two young men responded simultaneously by nodding their heads. The three passengers that had just arrived boarded the mini-bus taxi and occupied the seat directly behind the driver's seat, but I overhead the man saying "our youth have no respect", referring to the two young men. From their conversation, I could see that these three passengers knew each other and were together. As the older man sighed, the younger woman asked him if the knees were sore. He nodded and smiled as the older woman held the man's hand to comfort and reassure	6.20am – 6.22am

him as they looked into each other's eyes

Another passenger arrived and she was a young
Chinese lady probably in her late twenties and she
was carrying a huge backpack. She seemed very shy
and did not say a word to anyone including the two
young men who were still standing near the 6.33am-6.34am
entrance of the mini-bus taxi and the passengers
that had already boarded the mini-bus taxi. The
Chinese lady quickly boarded the mini-bus-taxi and
sat on the seat directly in-front of the backseat.

A short-while after she was seated, the Chinese lady
then looked in my direction and asked me in
English if I could please show her mini-bus taxis 6.35am
destined for Town Z once we get to Town Y. I
smiled and she smiled back as I said "yes, I would
show her"

All of a sudden, a flurry of approximately twelve
passengers all arrived at once at the entrance of the
mini-bus taxi we had boarded. They had probably
been just dropped by another mini-bus taxi from
another location. They shoved their way through
and quickly filled up the remaining nine passenger
seats. Unfortunately, out of this group that had just
arrived, two young African ladies and a middle aged 6.37am
African woman had to be left outside as the mini-
bus taxi driver closed the door because the mini-bus
taxi was already full. The two young African men
who had been standing near the entrance of the
mini-bus taxi for a while were unfortunately caught
off-guard by the group of passengers that had just
arrived, and as a result, they were also left behind
because there was no longer any vacant seat on the

mini-bus taxi. They tried to plead with the driver to please allow them to squeeze in as they were already late for work. Unfortunately the driver refused and told them that they had learnt a lesson, which means that they had to endure another long wait for the next mini-bus taxi to Town Y.	
As the mini-bus taxi departed, the mini-bus taxi driver requested the passengers to pay their fares. I noticed that the passengers already knew that they had to combine their fares according to their sitting order per row on the mini-bus taxi. All the money had to be pushed to the passenger sitting next to the driver who already knew that by merely sitting next to the driver she had to count all the money and provide change to those passengers who required change on behalf of the driver. Once all the money was collected, the passenger sitting next to the driver handed all the money to the driver who did not even say thank you	6.38am-6.40am
As the mini-bus taxi was on its way, a certain African man probably in his late seventies started a conversation with another African man also probably of the same age, and they were reminiscing about times gone by while they were still in their prime. They were talking so loud that everybody on the mini-bus taxi was listening to their interesting conversation, and everybody was laughing and enjoying the conversation. I saw most of the passengers looking at the Chinese lady as if they were surprised. I saw the Chinese lady getting worried and anxious.	6.40am-6.454am

One chose the mini-bus taxi because all people including all the different genders, age groups and backgrounds, have an equal chance of using the mini-bus taxi as it is the most popular mode of public transport in South Africa.

The interpretation of the older man and woman's unhappiness when the two young men nodded after enquiring about the mini-bus's destination is that the older man and woman expected the young men to show some respect to older people. There is a possibility that such youth were urbanized as they did not see anything wrong with just nodding their heads. There is also a possibility that the two young men were absent-minded or their minds were preoccupied with the thought of arriving late at work.

The Chinese lady who had boarded the mini-bus taxi seemed to be an unusual sight for most of the passengers, and the continued conversation in the Zulu language and the laughter made her feel not at ease. She was probably not sure if they were talking about her or laughing at her as she was the only person who looked different from everybody on the mini-bus taxi.

The interpretation of the driver's refusal to allow the two young men to squeeze in despite the fact that the mini-bus was already full was that the driver was a man of principle and put the safety of passengers first, hence he did not want to carry an overload of passengers on the mini-bus.

The interpretation of the way passengers enter the mini-bus taxis is that it is more like a first come first served basis or survival of the fittest. It is always best to be early and on-board immediately instead of standing outside because other passengers may overtake you. If the Chinese lady had arrived late together with the group of passengers who pushed and shoved their way into the mini-bus taxi, she could have panicked.

From the experience, one learned that it is possible to gain a rich insight on a natural environment as the people are not aware that they are being observed and how passengers interact on a mini-bus taxi.

One felt at ease conducting this research as one knew that the passengers did not notice that one was observing them, and sitting at the back seat also helped me to have a better vantage point.

The limitations of the exercise and of observing are that it may put the researcher's life in danger if those observed realize that they are being observed or make political assumptions about the researcher especially in unfamiliar settings (Green & Thorogood, 2009). If it is discovered that there is deceit, this could result in upset and premature termination of the research (Saunders et al, 2003). Moreover, the observer's prejudice may affect the validity of the research (Welman et al, 2005).

There could be ethical implications of carrying out this covert observation as the consent of the observed has not been sought and the autonomy of the observed is violated (Green & Thorogood, 2009). Covert observation necessarily implies that the subject is deceived or her right is infringed on (Mouton, 2002).

References

Green, T. & Thorogood, N. (2009) *Qualitative methods for health research*. London: Sage.

Herrera, C.D. (1999) 'Two arguments for covert methods in social research', *British Journal of Sociology*, 50(2), pp. 331-343.

Jefson, C.A. (2005) 'A review of the social construction of sexuality', *American Journal of Sexuality Education*, 1(1), pp. 183-188.

Kitchin, R. (2010) 'Out of place: knowing one's place: space, power and the exclusion of disabled people', *Disability & Society*, 13(3), pp. 343-356.

Larkin, M. (2011) *Social aspects of health, illness and healthcare*. Berkshire: McGraw-Hill.

Lerum, K. (2001) 'Subjects of desire: academic armor, intimate ethnography, and the production of critical knowledge', *Qualitative Inquiry*, 7(4), pp. 466-483 [Online]. Available from: http://qix.sagepub.com/content/7/4/466 (Accessed: 22 May 2013).

Liampuntong, P. & Ezzy, D. (2005) *Qualitative research methods*. New York: Oxford

Maree, K. (2010) *First steps in research*. Pretoria: Van Schaik.

McKenzie, J.S. (2009) 'Practices in an overt participant observation study', *Sociological Research Online*, 14(2). Available from: http://www.socresonline.org.uk/14/2/8.html

Moffat, S., White, M., Mackintosh, J. & Howel, D. (2006) 'Using quantitative and qualitative data in health services research – what happens when mixed method findings conflict?' *BMC Health Services Research*, 6, p. 28.

Mouton, J. (2002) *Understanding social research*. Pretoria: Van Schaik

O'Mahony, P. (2011) 'Social theory and the critique of capitalism in a communication society', *Irish Journal of Sociology*, 19(1), pp. 150-175.

Saunders, M., Lewis, P. & Thornhill, A. (2003) *Research methods for business students*. Harlow: Prentice Hall.

Spicker, P. (2011) 'Ethical covert research', *Sociology*, 45(1), pp. 118-133.

Ulin, P.R., Robinson, E.T. & Tolley, E.E. (2005) *Qualitative methods in public health*. San Francisco: Wiley.

Van Niekerk, A. A. & Kopelman, L.M. (2005) *Ethics & AIDS in Africa: the challenge to our thinking*. Claremont: davidphilip.

Welman, C., Kruger, F. & Mitchell, B. (2005) *Research methodology*. Cape Town: Oxford.

Chapter 3

Qualitative Research Methods

This chapter will cover the following areas:

- Interviews
- Focus groups
- Texts
- Diaries

3.1 Interviews, focus groups, texts and diaries

FOCUS GROUPS AND INTERVIEWS are some of the most commonly used methods of qualitative data collection (Joubert & Ehrlich, 2007). A focus group is defined as a qualitative data collection method in which a small group of approximately 6 to 12 people comprising one or two researchers (a moderator and a note-taker) and participants with similar social or cultural backgrounds or who have similar concerns, come together to discuss a particular research topic, and this is usually tape recorded or video-taped (Mack et al, 2005).

The aim of using focus groups is not to replace individual interviewing, but it is to gather insight and rich information that cannot be easily obtained through individual interviews (Welman et al, 2005). For instance focus groups have the unique feature of 'group effect' which is the interaction within the group which enables the eliciting of responses between participants in the group to produce insight and rich information that would not be accessible through individual interviews (Liampuntong & Ezzy, 2005). Green & Thorogood (2009), add that focus groups may be useful in social marketing and in gaining an insight about the perceptions of people to a certain health risk.

Other authors observe that focus groups have been applied increasingly to population and health research, and are useful for formative research and evaluation of outcomes (Ulin et al, 2005). It is therefore possible that focus groups could be a useful tool to gain an insight about the perceptions on male circumcision among men who have been circumcised as part of the male circumcision campaign aimed at reducing the spread of HIV in Eswatini.

It is important to highlight that a focus group is generally a small number of individuals or interviewees (ranging between 6 and 12 people) in that group for the purpose of collecting qualitative data (Ulin et al, 2005). Other authors suggest that a typical number should be eight to twelve, with a maximum of twelve (Mack et al, 2005). Liampuntong & Ezzy (2005), indicate that as a rule of thumb, there are about six to ten participants in one focus group session, but some sessions may have up to twelve people. The argument is that if the number is less than six it is less likely that discussions will express wider norms, values or opinions (Ulin et al, 2005).

On the other hand if the number of participants in the group is more than twelve, this may create problems because it is not easy to manage a large number of participants in a group, and some participants may find it difficult to talk in a big group where everyone is trying to talk (Liampuntong & Ezzy, 2005). For example, one has seen external service providers coming to collect qualitative data at a workplace, and it was a challenge for them to manage the group as they used a very large group of participants comprising men and women of various ages. Larger focus groups may also pose serious challenges for qualitative researchers that are still trying to find their feet in gaining a competence in the use of this means to collect qualitative data (Saunders et al, 2003).

Moreover, it is important to highlight that the focus group should be fairly homogenous especially where sensitive issues are discussed (Joubert & Ehrlich, 2007). For example the external service providers who employed focused groups at our workplace should not have mixed men and women in the same group as the issues of sexuality that were discussed proved to be a very sensitive topic.

A focus group is a useful qualitative data collection method that is likely to yield good results if it is restricted to between six and twelve participants. Anything less or more than that is bound to create challenges for the qualitative researcher.

There is argument about the importance of diversity in the composition of focus groups. Composition is one of the important variables in focus groups (Hoppe et al, 1995). As the emphasis is on group discussion, the composition of the group plays a major role in the interaction process (Liampuntong & Ezzy, 2005). Conventionally, it is argued that participants should have something in common so that maximum interaction within the group can be achieved and individuals dominating or withdrawing can be avoided. Generally, the three areas of concern regarding this commonality are: A homogenous group or heterogeneous group; shared experiences; and familiar faces or strangers (Liampuntong & Ezzy, 2005).

Many texts advocate having strangers in a focus group session. However, that line of argument is challenged because there are many situations when using strangers is not permissible or practical, as in these situations familiarity rather than anonymity may be the key to free-flowing discussions (Liampuntong & Ezzy, 2005). For example, in many parts of the rural areas of southern Africa, it may be a challenge to have older women and young girls in the same focus group to discuss openly about sensitive issues such as sexuality, but these women and girls may be open to discuss these issues with their acquaintances.

Natural groups must suffice when they are necessitated by pragmatic constraints or when the goals of the research require, or are enhanced by, the presence of a pre-existing group dynamic (Leask et al, 2001). It is recommended that focus groups of children be homogeneous in terms of gender and age as large discrepancies in age and gender can be problematic (Hoppe et al, 2005).

Saunders et al (2005) suggest using a horizontal slice through an organization to select a sample of people who have a similar status and similar work experiences to construct each group. Saunders et al (2005) add that using a vertical slice would introduce perceptions about status differences and variations in work experiences. In this way, group interviews can be conducted at a number of levels within an organization.

It can be said that the composition of focus groups has been an issue of debate among some researchers, and that some research questions are more suited to one type over the other. It is argued that careful consideration and common sense play a vital role in the selection of research participants in focus groups (Liamputtong & Ezzy, 2005).

Although the use of video has been a subject of debate among researchers, some researchers highlight that using video in qualitative research has some benefits in qualitative public health research. For example, some researchers observe that it is relatively easy to use, it is accurate, provides explicit and rich information about the research setting and interactions, which proves to be reliable (Lomax & Casey, 1998).

Moreover, the video-taped qualitative research setting and interactions gives the researcher an opportunity to review the information at a later stage, and the fact that video-taped information can be re-used and analysed by other researchers or for academic teaching purposes (Lomax & Casey, 1998). Green & Thorogood (2009) posit that using videos in qualitative research provides more possibilities such as the inclusion of hyperlinks with video clips in qualitative research articles.

The benefits of using diaries in conjunction with other methods in qualitative health research are that diaries can become a form of early data collection (Green & Thorogood). Diaries can also be of benefit in the sense that they can become a comparable and contextual data and can be useful for triangulation of research findings, and later reflections on what has been recorded in diaries can result in unforeseen discoveries (Saunders et al, 2003). Green & Thorogood (2009) highlight that using diaries in conjunction with interviews or observation may mean that there may be no need for interviews or extended observations in certain instances.

Diaries can be a powerful tool to complement other qualitative methods such as interviews and observations, and that using diaries in conjunction with interviews, does not negate the need for extended interviews and observations. However, it is important to appreciate that there are different types of diaries as diaries differ in depth and length (Valimaki et al 2007). Some authors mention four types of diaries which include meagre diaries, reporting diaries, descriptive and reflective diaries, and highlight that descriptive and reflective diaries are laden with rich information that they could be analysed in a similar way as interviews, which shows that descriptive and reflective diaries are more suited for qualitative analysis (Valimaki et al, 2007). This suggests that descriptive and reflective diaries could complement interviews.

Jacelon & Imperio (2005) are interesting when they mention two types of diaries such as solicited and unsolicited diaries, and highlight that unsolicited personal diaries are often used by biographical researchers to explore historical events. Green & Thorogood (2009) caution that diaries that are used to explore historical events should be analysed as literary constructs.

On the other hand, solicited diaries have a different dimension, and there have been suggestions to incorporate solicited participant diaries into data collection strategies for qualitative research, because solicited diaries are written with the researcher in mind as the participant is both the observer and informant (Jacelon & Imperio, 2005).

The combination of a solicited diary followed by an unstructured interview focused on the events chronicled in the diary is a good method of obtaining rich data (Jacelon & Imperio, 2005). Perhaps the observation that there may be no need for extended interviewing or observation when such diaries are combined with other qualitative research methods comes from that premise.

Solicited participant diaries can be incorporated into data collecting strategies for qualitative research, and this emphasizes the usefulness of diaries in complementing other methods such as interviews and participant observation. What is worth highlighting though is that for a researcher to use solicited diaries as a strategy for data collection, participants must have certain abilities such as literacy, ability to write in the language in which the research is conducted, and physical capacity (Jacelon & Imperio, 2005). This could really be a challenge in areas where there is low level of literacy.

The use of social marketing approaches in public health practice appears to be increasing, as social marketing has emerged as a recognizable practice. According to Wilson et al (2013), social marketing is an approach which targets specific audiences or target groups with marketing strategies to improve personal health and quality of life, and in general there are five major components of developing social marketing strategies which include the "5 Ps:" product, price, place, promotion, and positioning. Social marketing employs the application of commercial marketing tools and principles to the design, implementation, and evaluation of health and social behavior-change programmes (Ross et al, 2006).

Focus groups are key to understanding the target audience in a way that would not be accomplished with quantitative data alone (Burroughs et al, 2006). Focus groups have been widely used in social marketing because they allow the researcher to understand what people think, how they think and why they think as they do, and to examine how diverse identities and social networks might impact upon research participants' perceptions of a particular health issue such as HIV and AIDS and their reactions to the media coverage (Liampuntong & Ezzy, 2005).

The fact that focus groups have the unique feature of 'group effect' enhances discussions that could generate important insights into the price, place, promotion strategies, values and motivations that affect consumers' decisions to adopt the product. This could possibly guide the development of a social marketing programme to reduce risky behaviours. Other authors add that focus group data and community steering group involvement is critical for developing a grassroots social marketing strategy to improve health and risky behaviour (Wilson et al, 2013).

In social marketing the health promotion outcome may not be identical to the product or behavior being promoted, such as free-time physical activity among children by marketing "fun" and "cool" activities, rather than focusing primarily on health benefits or health risks (Tan et al , 2010). Social marketing is increasingly being seen as potentially effective means of pursuing health education practice generally and within specific areas as mental health and wellbeing and more broadly in tackling health inequalities. Social marketing-based public health interventions could be particularly useful in addressing health behaviours such as physical activity, where health education alone may not be effective (Tan et al, 2010).

Brief descriptions of the three interviewees that were chosen:

> All the chosen interviewees do not fall in the vulnerable category of people and they have the capacity and autonomy to make personal decisions as per ethical requirements (van Niekerk & Kopelman, 2005). They are:

- Chief Operations Manager: He was chosen because he is the most influential person in terms of the operations of the business, at the Port of XX.
- Full Time Labour Union Representative: He is the voice of all employees and has great influence.
- Communications Manager: She is influential in terms of being the voice of the company.

- Interview schedule:

- Each interviewee will be approached individually and a brief description of the research will be explained to them.
- Each interviewee will be requested to sign a consent form to acknowledge that they agree to participate in the research study and that they agree to be audiotaped during the interview.
- The interviewees will be interviewed individually in a private office with no interference.
    - It is planned that the Chief Operations Manager will be interviewed from 10am to 11am, Tuesday, 24 May 2022.

    - It is planned that the Full Time Labour Union Representative will be interviewed from 2pm to 3pm, Tuesday, 24 May 2022.
    - It is planned that the Communications Manager will be interviewed from 12 noon to 1pm, Tuesday, 24 May 2022.

- The 5 questions that each participant will be asked are:
    - Can you please tell me about what you understand about health?
        - Probe: How is health important to you?
    - What would you regard as the pressing health issue in your company?
        - Probe: How does this health issue affect the business?
    - What has the company done to address this health issue?
        - Probe: How effective has the company's health intervention been?
    - From your perspective, how would you ensure that employees are fit for duty?
        - Probe: What could prevent you from ensuring that employees are fit for duty?
    - How integral is the health of employees to the business's success or lack of?
        - Probe: Why is that so?

Example of Interviews

The research question is "How is health perceived by management and trade unions at the Port of XX?"

The questions one asked were sufficient, and they did obtain the information one wanted. One realized that one could not ask the questions in the same order for all the respondents to enable the interview to flow smoothly. One believes the flow of the discussion was smooth because it was not restricted to a particular order but was influenced by the direction the responses of each respondent took, as this was a semi-structured interview. Green & Thorogood (2009), report that there is flexibility of asking questions in semi-structured interviews.

As recommended by Liampuntong & Ezzy (2005), I recorded the information through a tape recorder and also took notes. However, one realized that it was not so easy to take notes and interview at the same time. The tape-recorded information helped me a lot because one could re-visit what actually transpired during the interviews from which one could produce the transcripts.

What one recalls about the interviews afterwards was that the first interview was good but there was an element of tension towards the end as the respondent was clearly in a hurry to finish as he was rushing to another meeting. But what stood out in that first interview was that the respondent appeared to be more concerned about the performance of the business rather than the health of the workers. The second interview went well, and the respondent was calm and seemed to enjoy the subject of the interview and expressed herself openly. The third interview also went well, and one can recall that the respondent was very emotional about the exploitation of workers and the health challenges facing workers. The third respondent was also able to look at things from a broader perspective as he was also concerned about the irresponsibility of some workers who did not take charge of their own health such as by not joining medical aid.

What one learned from this experience is that the respondents were so relaxed and were willing to share as much of their opinions and experiences no matter how sensitive it was. This showed that trust is very important when conducting interviews. According to Welman et al (2005), frankness and honesty are required from the interviewer to ensure that the respondent expresses himself/herself freely as he/she knows that her anonymity is also guaranteed. One also learned that as an interviewer, one does not need to say a lot, but must allow the respondent to express himself/herself without making him/her feel dominated or overwhelmed by me as the interviewer. This enabled one to gain a better insight into how the respondents perceived things from their own point of view. Mack et al (2005), highlight that such interviews have the value of making us able to probe so that we could be able to see the connections or associations between what the respondents say and particular phenomena.

Site: Executive Office – Port of XX

Interview C001

Interviewer: YYZ Mabuz

Transcriber: YYZ Mabuz

Date: 24 May 2022

Start: 10h00

End: 10h28

Question 1

Interviewee: Can you please tell me what you understand about health

Respondent: I don't think I can give you a doctor's definition, but I can give you the definition of health from an operations manager's perspective. Let's see ... well, well, health depends on what you are referring to. Right now I would say if you do not have any issues that bother you, then you are healthy, but if you are constantly worried, it means you are not healthy. Does that make sense?

Interviewer: is that all you can tell me about your understanding of health?

Respondent: Yep! That's it. But perhaps one could add another dimension and say, if you guys declare us fit for duty it means we are healthy, but if you don't , then we are in trouble, because that would mean there is something-something somewhere not right. We have seen your monthly reports indicating that there is a growing trend of employees with bp, diabetes, blah blah blah eehh () that our employees are not attending the clinic as they should.

Question 2

Interviewer: What would you regard as the pressing health issue in your company?

Respondent: Mmhh ... let me let me say ehh ... how can I put it? I know you guys are currently busy with the health awareness month sessions in the various departments of the company. I am sorry that I could not attend the sessions you had with the management team last Tuesday. You know, it is so hectic for me, cos I've been attending incident investigation after incident investigation during the past two months due to the frequent number of disabling injuries on duty that we are faced with. You know what... I'm so frustrated. By the way, what was your question? Oh...yah, I remember. But I think I've already answered your question because judging from the high disabling frequency rate in this company, it shows that we are not healthy, all of us here. We are running like headless chickens. Eish... this thing frustrates me. Ok, let's carry on, as I have to rush to another meeting eeh... at half past

Interviewer: How does this affect the business?

Respondent: The department of labour's eyes is on us. They are watching us very closely, which means that the company is in the black books. You can imagine. You can imagine for yourself, our clients will not be too pleased to do business with us because we are now seen as an unsafe working environment, especially for the past two months. Things have been really very bad for us. Losing our clients will not be good because the business has been going through a bad patch during the past twelve months. The volumes and tonnages are very low, and we were not able to reach our norms and stretch targets by far during the past financial year. This is so ironic because the business has been performing so brilliantly for so many years now, and there is already a huge budget set aside for further expansion of this port.

Question 3

Interviewer: If you regard this as the pressing health issue affecting the business, what has the company done to address this issue?

Respondent: Obviously we have the on-site clinic, and we have approved that the clinic operates 24 hours per day to ensure that emergencies that happen at night are promptly attended to. What else? ()

Interviewer: How effective has this intervention been?

Respondent: Mmmhh... () how can I put it ... (). It is a bit tricky to tell because the clinic has been operating 24 hours for only three months now. But I could say we have not yet seen that much of an impact because we still have a backlog of medicals as you always report at the daily MDT.

Question 4

Interviewer: From your perspective, how would you ensure that there is an impact to ensure that the backlog of medicals is addressed promptly and that employees are fit for duty?

Respondent: You got me there... ehh... ehh... This a bit tricky. What can I say (). It means we have to stop the operations to allow people to visit the clinic. You know, if we can release all the employees to go to the clinic at the same time, there will be huge delays in operations as our clients also demand a quick turnaround. Time is of the essence here because any delay in operations will come back to haunt us as we will not be able to meet our norms and targets, and we may even incur huge penalties.

Interviewer: Ehhe ...

Respondent: If the business does not perform, my head is on the block. We are talking business here. There is a lot of pressure from clients and from head office. I do not want to be seen as a failure. At the end of the financial year, everybody expects to get a bonus, but they do not understand that it is through the hard work we put in operations () ... that makes the bonus possible.

Question 5

Interviewer: How integral is the health of employees to the success of the business or lack of?

Respondent: That is why we have you and the clinic to ensure that you support us as a business in that aspect.

Interviewer: Please elaborate

Respondent: I mean to say ... () that is your competence area as health professionals to look after the health of our employees. That is why we employed you to look after that aspect. You know! I am running late for the next meeting, please excuse me.

Interviewer: Thank you for your time Sir.

Respondent: Sharp Doc ... ().

Site: Corporate Affairs Office – Port of XX

Interview C002

Interviewer: YYZ Mabuz

Transcriber: YYZ Mabuz

Date: 24 May 2022

Start: 12h17

End: 12h53

Question 1

Interviewer: As the person in charge of communications in the business, what is your understanding of health?

Respondent: To me health is a very broad term. I have learned a lot from doing my own research as you know that our portfolio requires us to ensure that we communicate correct information that has been verified to ensure that we do not send wrong messages that will create problems and confuse people, and to avoid tarnishing the name of the company. On the one hand, health means being at peace with yourself and on the other hand health means being physically well, and ensuring that you eat well, you sleep well, you have regular medical check-ups to ensure that there is nothing wrong with you. I guess it depends from which angle you look at it. There could be many definitions. But as far as I am concerned that is what I can describe health as.

Interviewer: Okay ...

Respondent: Just one thing that comes to my mind ... () I remember last year, I thought I was healthy because I thought I was coping with the pressure at work. And all of a sudden, boom! I was hospitalized. It is only then that I realized that I had burned both ends of my candle. Anyway, I was still new in this job and was trying to prove a point that I could make an immediate impact.... ((Laughs loudly)). The hospitalization brought me down to reality. I wish we could all realise how vulnerable we are. I value my health so much now. It was such an unpleasant wake-up call when I got hospitalized, and that was my very first time ever to be admitted in hospital!

Question 2

Interviewer: What would you regard as the pressing health issue in your company right now?

Respondent: Mmhh (...) it's quite (..) interesting that you are asking me this question because it was highlighted at our EXCO meeting yesterday that absenteeism is getting out of hand at this place. Judging from the trends that were projected, I can conclude that our workforce is really sick. I have already put these absenteeism trends on the plasma screens and loo doors to make everybody in the business aware of the huge challenge that we face. Everybody must come on board and play a part.

Interviewer: Ehhe

Respondent: (...) the health communication partnership between the health department and our department has opened my eyes big time! I never imagined that high blood pressure and diabetes were so serious in this company. At least we are doing our best to communicate this message in all available platforms of the company. I am sure the change champions that have recently been launched will play a key role in changing the culture. It is evident that things have been taken for granted for too long in this company, specifically matters that touch on health (...) We as individuals as well, and I am testimony to that.

Question 3

Interviewer: In your opinion, how integral is the health of employees to the performance of the business and why?

Respondent: It is very very integral because if you look at the high absenteeism trends that I have just mentioned, it means that the operations of the business is compromised when workers are absent. () I believe that if workers are healthy and not frequently absent from work, the business can do very well.

Interviewer: Please elaborate

Respondent: Sure (). The crux of the matter is that if you look at it on a broader scope, if people are unhealthy whether they come to work or not the performance of the business gets affected (...) Do you get my point? If you are sick and you are at work your performance is not 100 per cent. If your performance is not 100 per cent, it is obvious that the performance of the business will also not be 100 per cent. (...) obviously, when you are absent the business suffers because there is no one to perform your duties, even if they find a temporary replacement, that temporary replacement is likely to make mistakes. I have seen that even in my department whereby two of my workers were frequently absent last year, and this really hit us big time because I had to perform their duties as well as my duties, but I realized that some of the duties had to be compromised over others. I had to take a stand and recruit interns to understudy these frequently absent employees, to ensure that the work carries on without much of a negative impact when one or both of my workers are absent. If we could feel the negative impact as a service department, how much more if the absenteeism happens in operations where the bread and butter of our business lies. It is much deeper, very much deeper (). Coming here has opened my eyes to things that I used to take for granted during my previous job as a freelance journalist. And (...). Okay okay

Question 4

Interviewer: You have mentioned that the company is faced with the challenge of a high absenteeism rate due to the fact that the workforce is really sick. What has the company done to address this challenge of sick employees?

Respondent: The company has implemented the return to work interviews for every employee who returns to work after having been off sick. I have had a number of the return to work interviews with my two employees that were frequently off sick last year. What else has the company done (...) Yes, the company has also made it compulsory for us as line managers to report on absenteeism on a weekly basis, and there have been absenteeism training for all line managers and supervisors.

Interviewer: Do you think these interventions have been effective, and why?

Respondent: To a large extent, I would say no because when you look at the absenteeism trends, the graphs that we have projected on the plasma screens throughout the terminal, the graphs show an increase in absenteeism instead of a decrease. This shows that our employees are getting sicker and sicker.

Question 5

Interviewer: What would you suggest or recommend as the most effective way of addressing the health issues affecting your company?

Respondent: Ye ye ye (...) it is a tough one (...) ((shaking her head)). Let me be controversial and suggest that the company employs more women and younger people as my general observation is that there are very very few women in this company and the majority of the current employees seem to be old. It was pretty obvious at the long service awards event held last week that we have a lot of workers who have worked for far too long here and their bodies are tired now. Where are the women? We need more of the women ((laughing loud))

Interviewer: Okay

Respondent: Honestly, I think that is an area that needs to be looked into seriously. Something needs to be done very urgently

Site: Corporate Affairs Office – Port of XX

Interview C003

Interviewer: YYZ Mabuz

Transcriber: YYZ Mabuz

Date: 24 May 2022

Start: 14h04

End: 14h33

Question 1

Interviewer: As a full time representative of workers, what is your understanding of health?

Respondent: Thanks Doc, this is a profound and important question. Health is very close to my heart. My thinking is that health it is not easy to be completely healthy, and I am sure there is no single person in the world who can claim to be completely healthy. Having said that, I think it is important to distinguish things and compare apples to apples. There is a healthy person and there is a healthy environment. By healthy person I would say, that particular person has no hindrances that affect his or her mind and her body. A healthy environment is an environment where all people live and work in harmony, and there are no hindrances of any kind if we are to consider an environment to be healthy. In an environment such as this one at the port, I wouldn't regard it as a healthy environment because workers work under difficult conditions such as the dust, and harsh weather conditions. At the same time, the workers work very hard but the money is very little. Let me tell you a little story. (...) I remember when there was this long strike in 2010, when management had to wear overalls and did the manual work that the majority of our workers do on a daily basis, and they also felt how it is to be exposed to harsh weather conditions. Within a day, those managers could not take it any-more and some of them had to be admitted into hospital because they felt the strain of being exposed to unhealthy environmental conditions and of hard labour.

Interviewer: Okay

Respondent: It breaks my heart when I see our own African people who hold senior positions such as management disregarding the wellbeing of our fellow brothers and sisters, mothers and fathers, and fellow human beings in general. That is why our trade union is calling for the end of labour brokers because these labour brokers perpetuate the plight of our workers. In a nutshell, health to me means the absence of any form of exploitation.

Question 2

Interviewer: In your view, what would you regard as the most pressing health issue affecting this company that you work for?

Respondent: (...) Exploitation of workers, of course is a pressing health issue here. We may see other issues such as HIV and so forth and so forth, but the underlying factor is what makes people more sick is the environment and the conditions in which they live and work. I know that some people may dispute that, especially those that occupy management positions (...). But I maintain that point of view because it is the truth. The truth is painful, but I say it as it is. Our management knows deep down in their hearts that they are not in any way committed to making sure that the workers are healthy. Things are getting out of hand

Interviewer: Ehhe

Respondent: I must admit though that it would be unfair to shoulder all the blame on management, because we as workers also need to take responsibility of our own lives. What we eat and how we behave can be a factor on whether we feel healthy or not. At least being a change champion has helped me to look at things from a broader angle. Some people are just negligent of their own lives. They do not want to check their health status, and even if they have checked and been told that there is a problem they continue living as if nothing was wrong. I guess you have a huge load on your shoulders as it is not easy to change people.

Question 3

Interviewer: Exactly what has the company done to address health issues at the workplace?

Respondent: I must say that all workers are required to do annual medicals at the company's clinic. I am sure this has somewhat helped to pick up health conditions and to make people aware that they have such and such a condition. There are education and awareness sessions done by the health team on a daily basis. Right now the company has launched a hotline to assist workers and their family members who need assistance or advice on health issues and also a counseling service. You see, I am not just talking as a labour union representative, but am wearing both my hats as a labour union representative and as a change champion. That is fair, is it not? (...)

Interviewer: Ehhe

Respondent: most of these interventions are still relatively new, and therefore it is not that easy to gauge their effectiveness at this stage. These things need the continuous full support of the powers that be. Otherwise, they just will fall flat. We are doing our bit as change champions to support the cause

Question 4

Interviewer: In your opinion, how integral is the health of employees to the performance of the business and why?

Respondent: Employees are like the pillar of the company. So, if the pillar is weak or if the employees are not healthy, the company will also be weak, but if the pillar is strong and the employees are healthy, then the performance of the business will also be strong. Yes, this shows how integral the health of employees is to the performance of the business.

Interviewer: Okay

Respondent: (...) if a company does nothing to address the issue of health of its workers, such a company is doomed to failure. Workers health is so central to any business performance, no matter how small the company may be. In our case, the port is so huge, which means that health of workers is even more crucial.

Question 5

Interviewer: In your opinion, what would be the best recommendation you can give to your company to ensure that all

employees are fit for duty?
Respondent: (...) (...) ((scratching his head)) the best recommendation I can give is regular check-ups at the clinic. Oh! The other important thing is that the company must make it compulsory for all workers to join medical aid.
Interviewer: Okay
Respondent: The issue of medical aid is so crucial because if people do not belong to a medical aid they will have to face long queues in the state hospitals which do not often give a good service because they are so overcrowded. Right now, I went to visit one of our workers who is in a critical condition in one of the public hospitals and his family is so frustrated because the person has been lying there for almost three weeks and they were told that he is still waiting to do a scan. When we investigated, we discovered that the scan has been out of order since last year. It is such a crisis and drama in the state hospitals. So, I would emphasise that every worker should be on medical aid.
Interviewer: Thank you for your time Sir
Respondent: Eh (...) Thank you for the discussion, this issue makes me so emotional (...).

References

Burroughs, E.L., Peck, L.E., Sharpe, P.A., Granner, M.L., Bryant, C.A. & Fields, R. (2006) 'Using focus groups in the consumer research phase of a social marketing program to promote moderate-intensity physical activity and walking trail use in Sumter County, South Carolina', *Preventing Chronic Disease*, 3(1).

Coyne, I.T. (1997) 'Sampling in qualitative research. Purposeful and theoretical sampling; merging or clear boundaries', *Journal of Advanced Nursing*, 26, pp. 623-630.

Green, T. & Thorogood, N. (2009) *Qualitative methods for health research*. London: Sage.

Hoppe, M., Wells, E.A., Morrison, D.M., Gillmore, M.R. & Wilsdon, A. (1995) 'Using focus groups to discuss sensitive topics with children', *Evaluation Review*, 19(1), pp. 102-114.

Jacelon, C.S. & Imperio, K. (2005) 'Participant diaries as a source of data in research with older adults', *Qualitative Health Research*, 15(7), pp. 991-997.

Joubert, G. & Ehrlich, R. (2007) *Epidemiology: a research manual for South Africa*. Cape Town: Oxford.

Leask, J., Hawe, P. & Chapman, S. (2001) 'Focus group composition: a comparison between natural and constructed groups', *Australian and New Zealand Journal of Public Health*, 25(2), pp. 152-154.

Liampuntong, P. & Ezzy, D. (2005) *Qualitative research methods*. New York: Oxford

Lomax, H. & Casey, N. (1998) 'Recording social life: reflexivity and video methodology', *Sociological Research Online*, 3(2).

Mack, N., Woodsong, C., MacQueen, K.M. Guest, G. & Namey, E. (2005) *Qualitative research methods: a data collector's field guide* [Online]. North Carolina: Family Health International.

Ross, G., McDermott, L., Martine, S. & Kathryn, A. (2006) ' The effectiveness of social marketing interventions for health improvement: what's the evidence?', *Public Health*, 120(12), pp. 1133-1139.

Saunders, M., Lewis, P. & Thornhill, A. (2003) *Research methods for business students*. Harlow: Prentice Hall.

Tan, E.J., Tanner, E.K., Seeman, T., Xue, Q., Rebok, G.W., Frick, K.D., Carlson, M.C., Wang, T., Piferi, R.L., McGill, S., Whitfield, K.E. & Fried, L. (2010) 'Marketing public health through older adult volunteering: experience corps as a social marketing intervention', American Journal of Public Health, 100(10), pp. 727-734.

Ulin, P.R., Robinson, E.T. & Tolley, E.E. (2005) *Qualitative methods in public health*. San Francisco: Wiley

Van Niekerk, A. A. & Kopelman, L.M. (2005) *Ethics & AIDS in Africa: the challenge to our thinking*. Claremont: davidphilip.

Valimaki, T., Vehvilainen-Julkunen, K. & Pietila, A. (2007) 'Diaries as research data in a study on family caregivers of people with alzheimer's disease: methodological issues', *Journal of Advanced Nursing*, 59(1), pp. 68-76.

Welman, C., Kruger, F. & Mitchell, B. (2005) *Research methodology*. Cape Town: Oxford.

Wilson, D.K., George, S., Trumpeter, N.N., Coulson, S.M., Griffin, S.F., Wandersman, A., Forthofer, M., Gadson, B. & Brown, P.V. (2013) 'Qualitative developmental research among low income African American adults to inform a social marketing campaign for walking',

International Journal of Behavioral Nutrition and Physical Activity, 10(33).

Chapter 4

Participatory and Action Research

Participatory research is not a specific method but rather the use of multiple methods to solve a problem through group action (Ulin et al, 2005). It is a strength that the emphasis of participatory action research is on producing knowledge from the point of view of the marginalized, oppressed and deprived groups and classes as this has the potential of transforming social realities in order to improve the lives of all people (Liampuntong & Ezzy, 2005).

It is also a strength that participatory action research attempts to combine research, education, and action in one endeavour because through this people can take action for their emancipation, personal growth and change from their own perspective and not from the perspective of the researcher (Liampuntong & Ezzy, 2005). This enhances the strengthening of the local people's confidence and empowerment as ordinary people are enabled to validate the data used.

Not only is participatory action research valuable in developing countries, it is also a valuable approach in many developed countries especially where there are wide gaps between those who are privileged and those who are not (Liampuntong & Ezzy, 2005).

Just like many other types of research methods, participatory action research has its weaknesses. One of the potential complexities of participatory action research is that triangulation could be a challenge because the researcher may be required to cross-check the data with a wide range of people, given the complex social constructs (Liampuntong & Ezzy, 2005).

Since in participatory action research there is shared ownership between the researcher and the local people, publication of the data could be a challenge because the local people need to be clear from the start about why and when the data should be published (Liampuntong & Ezzy, 2005). Otherwise, if the local people have not been informed from the start, it may be a challenge to publish the findings. In this regard, it is highlighted that social action and empowerment is more important than publishing the findings (Ulin et al, 2005).

Participatory action research can be daunting and time-consuming as it requires involvement of the local people in all the stages of the research (Liampuntong & Ezzy, 2005). Unfortunately, funding bodies may not be willing to sponsor such projects, because in addition to the fact that they are time-consuming, they are also considered unscientific from the perspective of the dominant positivist paradigm, which could also compromise the chances of the findings being accepted for publishing (Liampuntong & Ezzy, 2005)

Since an authentic attitude of mutuality and openness and a commitment to learning may be required on the part of all people involved, particularly the researchers, this may present challenges if some people have other political agendas that prevent them from meeting these requirements (Liampuntong & Ezzy, 2005).

A participatory approach might be important in health research because it addresses the gap and perception towards health between health professional and lay people, and the realization that there are many factors which cannot be measured in biomedical terms. Therefore, the role of the researcher in participatory approaches is to have a shared responsibility with the local people rather than conduct research on them (Green & Thorogood, 2009).

There may be contradictory views on participatory research. A number of authors actually acknowledge that it is not straightforward to pinpoint the exact meaning of participatory action research as it means different things to different people (Liampuntong & Ezzy, 2005). It is very profound and just that "in principle participatory action research aims to examine the political structures that disempower marginalized, deprived, and oppressed groups of people and to find ways in which these structures can be changed" (Liampuntong & Ezzy, 2005).

There could be a lot of benefit to the communities and to the field of public health in general if most research was done using the participatory action research approach because this approach allows the participants to be co-partners rather than being imposed upon by the researcher. Green & Thorogood (2009) observe that in participatory action research as the community forms a collaborative partnership with the researcher, they also develop a reflexive awareness. Such collaboration could ensure that there is sustainability of whatever health intervention is implemented as a product of the research, as the communities will take ownership. The unfortunate reality is that most research that is conducted in developing countries does not involve the communities as genuine partners, and as a result, most of the health interventions that are implemented leading from the research become unsustainable.

There is a possibility that most of the public health research that is conducted in low income countries is more of an information extraction from the communities rather than a genuine partnership between the communities and the researchers. We have seen a number of internationally funded organisations that have conducted HIV and AIDS research in sub-Saharan Africa in the past decade or so. However, one wonders if there has been a genuine partnership between these research organisations and the communities in which the research is conducted. Some researchers argue if public health researchers and agencies can reconcile the push from funding bodies and the pull from communities (Green & Mercer, 2001). Perhaps, there could be some benefit if some rapport has already been established between the researchers and the community, but it has to be borne in mind that engagements to form genuine partnerships could be sensitive and may take time and energy (Camfield et al, 2009).

It is such a pity to realize that while qualitative research has received much attention in the past few decades, participatory action research as a concept and its practices has not received much attention in professional journals (Liampuntong & Ezzy, 2005).

In principle, participatory research seems attractive as it identifies with collaboration between the researchers and the community, but its meaning may differ depending on the context in which it is applied as communities' experiences also differ.

Some scholars have made an interesting point about participatory approaches being a 'tokenism or cosmetic labelling – to look good' in many cases. This is an important observation because there has been a dramatic increase in the attention and resources devoted to partnership or collaborative approaches to public health, which includes a call for greater community participation and control in the processes by which community problems and interventions are designed and implemented (Lanz et al, 2001).

It appears as if participation has become the buzz word in community research and development, and yet this is done to fulfil the agenda of the researcher or research sponsors rather than to truly emancipate the 'poor', the 'marginalised' and the 'oppressed'. Feminist researchers and action groups in many countries have questioned the terminology of the 'poor', the 'marginalised', and the 'oppressed' used in participatory action research (Liampuntong & Ezzy, 2005). It is argued that there is a danger in the use of categories such as the 'poor', the 'marginalised', and the 'oppressed' because this may imply that the 'poor', the 'marginalised', and the 'oppressed' are homogeneous (Liampuntong &

Ezzy, 2005). Feminisms argue that in practice and in theory, participatory approaches have often ignored the gender factor in oppression (Liampuntong & Ezzy, 2005)

If participatory approaches in health research that are conducted in a country which has high socioeconomic inequalities, should not be oblivious of the fact that both the community members and the researchers should equally be reflective of how and why they experience the world as they do. This is important, because, sometimes, it may be thought that the participatory research has empowered the community to take care of their own health, only to find that that would not be sustainable. If the health system is weak, it may be a challenge to have a community that will be able to effectively take care of their own health.

According to Giachello et al (2003), researchers and clinicians need to remember that without system reform, many communities cannot be able to adopt healthy lifestyle practices because options are simply not available. De Koning & Martin (1996), highlight the importance of critical reflection, an attitude of mutuality on all those involved in the participatory research, and warn the researcher to leave the position of those who have all the answers. Participatory approaches may look good in theory, but in practice, there could be many challenges.

Involvement of the community through all the processes in the research can result in their empowerment. Participatory approaches such as participatory action research prevent oppressed people from becoming passive objects who have research done on them, as in conventional research methods, because they are able to have equal power in controlling the research (Liampuntong & Ezzy, 2005). Trust between the researcher and the community is one of the most important ingredients in participatory approaches. During one's studies at the University of Stellenbosch's Africa Centre for HIV and AIDS, one observed that many of the participatory research that were conducted by non-governmental organisations and other institutions in the informal settlements of the city of Cape Town, failed to achieve their desired goal of empowering the communities. Perhaps, the communities were suspicious of exploitation by researchers, especially by those that looked different from them.

Leung et al (2004) observes that such distrust can be addressed by having communities as full partners in the research process, beginning with community identification of an issue. O'Fallon & Dearry (2002) emphasise the importance of communication for achieving benefits in community participatory approaches, and the benefits include trust between researchers and the community, which leads to increased relevance of research question, which leads to increased data quality and quantity, which leads to increased use and relevance of data, which leads to increased dissemination of findings, which leads to translation of research into policy, which leads to emergence of new research questions, which leads to the extension of research and intervention beyond specific project, which leads to infrastructure development and sustainability. Shared interests between researcher and community are therefore essential in conducting a good quality participatory research (Liamputtong & Ezzy, 2005).

There is complexity of the researcher's role in community participatory research. The key difference between participatory and conventional methodologies lies in the location of power in the research process. According to Ulin et al (2005), the researcher's skill as a facilitator-trainer may be more important for the project's success than her credentials as a research scientist. The researcher may have to make compromises on the choice of methods, arbitrate differences among participants on interpretation of the data, or find alternative solutions to a problem when the group's recommendations are inconsistent with available resources (Ulin et al, 2005).

Participatory approaches can empower communities and can also result in many other benefits for both the community and the researchers, but it is important that communication remains a constant throughout the entire spectrum of the participatory research process. The researcher's skill as a facilitator-trainer may be crucial for the success of the project.

Participatory approaches have become popular substitutes of the traditional top-down research approaches, indicating that the top-down research approaches have not really benefited the communities as much as they have benefited the researchers. The importance of contextualizing the research to the situation of the local population and the empowerment of the community should be highlighted. According to Green & Thorogood (2009), "the emancipatory aims of participatory approaches are perhaps more associated with research in developing country settings, whereas the collaborative and action orientated elements of these approaches have influenced healthcare research in developed country settings as well".

However, it has to be borne in mind that there could be a broader political agenda in participatory research of which one of the partners is unaware (Boutilier et al, 1997). For example, in the context of certain countries, the researcher may be interested in empowerment for social change, whereas, the local populations are interested in service delivery, as we often see in the ongoing service delivery protests that are so widespread in some countries. This tells us that a participatory approach in this regard may have negative implications for the research project, as the aims of the researcher and the local population may not be met, due to the different political agendas between the researchers and the local population.

Much as the emancipatory benefits of participation have been highlighted, it is also possible that some local populations may choose not to participate at all, no matter how well the researchers may have negotiated for the full participation of the local population. Interestingly, some researchers argue that the choice not to participate can actually be viewed as an act of empowerment, supposedly only achieved at the higher end of participation (Hayward et al, 2004). It is therefore observed that assessing social exclusion by measuring levels of participation may be misleading and may not account for community members who have made the rational choice not to participate for a number of reasons (Hayward et al, 2004).

Participatory approaches could play a significant role in the emancipation and empowerment of local communities. However, it should be borne in mind that participatory approaches could also be used to fulfil a certain political agenda which could have implications for the research project as a whole.

REFERENCES

Boutlier, M, Mason, R. Rootman, I. (1997) 'Community action and reflective practice in health promotion research', *Health Promotion International*, 12(1), pp. 69-78.

Camfield, L., Crivello, G. & Woodhead, M. (2009) 'Wellbeing research in developing countries: reviewing the role of qualitative methods', *Social Indicators Research*, 90, pp. 5-31.

De Koning, K. & Martin, K. (1996) *Participatory research in health: issues and experiences*. London: ZED Books.

Giachello, A.L., Arrom, J.O., Davis, M., Sayad, J., Ramirez, D., Nandi, C. & Ramos, C. (2003) 'Reducing diabetes health disparities through community-based participatory action research: the Chicago southeast diabetes community action coalition', *Public Health Reports*, 118, pp. 309-323.

Green, L.W. & Mercer, S.L. (2001) 'Can public health researchers and agencies reconcile the push from funding bodies and the pull from communities', American Journal of Public Health, 91 (12), pp. 1926-1943.

Green, T. & Thorogood, N. (2009) *Qualitative methods for health research*. London: Sage.

Hayward, C., Simpson, L. & Wood, L. (2004) 'Still left out in the cold: problematizing participatory research and development', *Sociologia Ruralis*, 44(1), pp. 95-108.

Lanz, P.M., Viruell-Fuentes, E., Israel, B.A., Softley, D. & Guzman, R. (2001) 'Can communities and academia work together on public health research? Evaluation results from a community-based participatory research partnership in Detroit', *Journal of Urban Health: Bulletin of the New York Academy of Medicine*, 78(3), pp. 495-507.

Leung, M.W., Yen, I.H. & Minkler, M. (2004) 'Community-based participatory research" a promising approach for increasing epidemiology's relevance in the 21st century', *International Journal of Epidemiology*, 33(3), pp. 499-506.

Liampuntong, P. & Ezzy, D. (2005) *Qualitative research methods*. New York: OxfordUlin, P.R., Robinson, E.T. & Tolley, E.E. (2005) *Qualitative methods in public health*. San Francisco: Wiley

O'Fallon, L.R. & Dearry, A. (2002) 'Community-based participatory research as a tool to advance environmental health sciences', *Environmental Health Perspectives*, 110(2), pp. 155-159.

Ulin, P.R., Robinson, E.T. & Tolley, E.E. (2005) *Qualitative methods in public health*. San Francisco: Wiley

Chapter 5

Data Analysis

This chapter will focus on the following topics:

- Introduction to data analysis
- Types of qualitative data analysis

5.1 Introduction to Data Analysis

THE TERM "ANALYSE" means to separate into component parts, so that these component parts can be individually understood and also understood in the context of the bigger picture. According to Jooste (2009), to analyse is to find patterns and relationships, to make inferences and to state the problem. Ulin et al (2005) explain that in social research whether qualitative or quantitative, analysis implies that the researcher systematically examines data to discover patterns. In other ways, qualitative and quantitative data analysis diverge such that in quantitative studies researchers identify key explanatory variables and expected outcomes in advance, whereas in qualitative studies researchers are interested in exploring the broader psychological, social, political, or economic contexts in which research questions are situated (Ulin et al, 2005).

It is important to analyse qualitative data and not to simply describe it because there is a growing expectation in the field of health research that qualitative studies must show transparency of the data analysis in order to improve rigour, reliability and validity (Green & Thorogood, 2009). Pope et al (2000) observe that the application of interim analysis at the data collection stage of qualitative research, has the advantage of allowing the researcher to go back and refine certain aspects of the research question as new patterns emerge, and it also allows the researcher to look for deviant cases as the emerging patterns of enquiry are examined in depth. Ulin et al (2005) add that field analysis can also lead to inclusion of new participants and looking at the data after every observation, interview or focus group session keeps the researcher alert to new discoveries and the potential for modifying the research process to strengthen the data. The analysis still carries on even after the researcher has left the field which allows the researcher to re-examine the data in depth and to systematically arrange it into categories and the important thing to note is that such continuous analysis in qualitative research is all inclusive (Pope et al, 2000).

Interpretation is part and parcel of analysis. Put simply, analysis is like identifying and displaying all the different pieces of a puzzle, and interpretation is like finding ways in which these different pieces of the a puzzle fit together and to derive meaning from the picture that emerges. Ulin et al (2005), mention that the data analysis process in qualitative research includes a sequence of interrelated steps which include reading and re-reading each set of notes or transcripts for content, coding or identifying the emerging themes, displaying the data or distinguishing nuances of a topic, and interpreting. At each step, the researcher searches for the core meaning of the thoughts, feelings and behaviours described in the texts – that is, the researcher actually interprets the data (Ulin et al, 2005). An overall interpretation helps to see how the emerging concepts and themes relate to the original research question and how the findings of the study can be applied beyond the specific context of the study (Ulin et al, 2005).

Qualitative analysis is a very detailed process and it requires to be systematically approached with an element of focus and imagination. Not all qualitative interpretations can be considered equal or accurate (Liamputong & Ezzy, 2005).

Continuous analysis in qualitative research is all inclusive. This means that, unlike in quantitative research, where the data is reduced to a few numerical codes, analysis in qualitative research involves constant comparison of all the data through a coherent and systematic approach which includes the addition of categories to reflect as many of the nuances of the data as possible (Pope et al, 2000). In fact, such inclusiveness is regarded as a key point in the process of constant comparison when analyzing qualitative data (Pope et al, 2000). The process of constant comparison, cross-checking and rigorous reflexivity also constitutes rigour in the context of qualitative research. Within this context, the onus is on the researcher to deal with the participants with integrity, honesty and fairness (Liampuntong & Ezzy, 2005).

The process of constant comparison could yield massive amounts of data, and managing the process could be laborious, which could pose a challenge not only to a novice researcher, but to established researchers as well. Perhaps, that is why certain software has been designed to manage such laborious data analysis. La Pelle (2004) highlights that using general purpose software tools can be used for analysing and managing many kinds of data, including key informant interviews, focus groups, document and literature reviews, and open-ended survey questions. At the same time, since the software is capable of handling large amounts of data, it is possible that some researchers may end up using larger samples with the hope that this would increase the statistical power of their research.

For example, a former colleague of ours was sharing his experiences of a qualitative study that he had conducted a few years ago in the South African Province of the Western Cape. He said his supervisor had told him to select a larger sample with the hope that the analysis would be easily managed through the computer software. Later on, his supervisor realized that they had made a mistake by selecting a larger sample, because the software analysis worked poorly for the task at hand. It is therefore not surprising that some authors caution that larger samples are really not always necessary in qualitative research as the sample size in qualitative research should be directed by the research question and analytical requirements, such as data saturation, rather than by available software (Pope et al, 2000).

Qualitative data analysis is inclusive because it involves the process of constant comparison which includes the addition of categories to reflect as many of the nuances of the data as possible as opposed to just reducing the data to a few numerical codes. This process of qualitative data analysis or constant comparison is systematic, rigorous, laborious and time-consuming. Software options available for qualitative data analysis can be good news and bad news (Ulin et al, 2005).

5.2 Types of qualitative data analysis

THEMATIC ANALYSIS IS the most basic and most common style of qualitative analysis, as it is used to establish categories of the common themes (Green & Thorogood, 2009). Joubert & Erlich (2007) elaborate that in thematic analysis, the data are explored in detail for common themes and these are then established into units of meaning or codes, and over time this process can be more complex as codes or categories may start to overlap, which may imply that sub-dimensions of these codes or categories may need to be developed.

Grounded theory is one of the more influential and deeper research methodologies in qualitative research, and it is consistent with the provisional and inductive approach of much of qualitative research (Liampuntong & Ezzy, 2005). In contrast to deductively derived propositions, grounded theory is designed to build an explanation or to carefully generate a theory around the core or central theme that emerges from your data (Saunders et al, 2003). The grounded theory approach is generally intense and time-consuming as it is structured and systematic, with set procedures to follow at each stage of analysis, and theoretical insights are constantly developed through constant comparison until the point of saturation (Green & Thorogood, 2009). In the grounded theory approach, data is fractured or disaggregated into units through a process called open coding, relationships between categories are recognized through the process known as axial coding, and the categories are integrated to produce a theory through the process of selective coding (Saunders et al, 2003).

Framework analysis is another deeper research methodology in qualitative research, and its ultimate goal is policy development (Green & Thorogood, 2009). Framework analysis generally involves a number of steps which include familiarization with the data in its entirety, thematic analysis to develop a coding scheme, indexing to systematically apply codes to the entire data set, and charting which involves arranging the data according to thematic content (Green &Thorogood, 2009).

One can see that thematic analysis and framework analysis appear to have something in common except that framework analysis takes a step further through the stage of charting which leads to mapping and interpretation or looking at the interrelationships between codes (Green & Thorogood, 2009). The key difference between framework analysis and the grounded theory is that in grounded theory the integrity of the accounts of the individual respondents is somewhat lost due to fracturing of the data, whereas in framework analysis the integrity of the individual respondents' accounts is preserved throughout the analysis (Green & Thorogood, 2009).

It would be important to consider the implications of using a particular approach before one commits to it. For example, one has observed that in a number of cases, it is difficult to get funding for qualitative research, and rather more so for an intense and time-consuming grounded theory approach, as there may be a concern that little of significance will emerge at the end of the research process.

Thematic analysis is the most common and most basic style of qualitative analysis, and may be sufficient for most exploratory qualitative studies. Deeper approaches such as grounded theory or framework analysis may be considered if there is a real need to move beyond what thematic analysis can provide (Green & Thorogood, 2009).

In the grounded theory approach, data is fractured or disaggregated into units through a process called open coding, relationships between categories are recognized through the process known as axial coding, and the categories are integrated to produce a theory through the process of selective coding (Saunders et al, 2003).

Open coding is essentially the first stage of the process of grounded theory analysis, whereby the data that is collected is disaggregated into conceptual units and provided with a label, and the same label will be given to similar units of data (Saunders et al, 2003). Green & Thorogood (2009) regard this stage as intense because it requires line by line analysis to open up or fracture the data. Moreover, because this research process commences without an explicit basis in existing theory, the result may be the creation of a multitude of conceptual labels or codes related to the lower level of focus and structure with which you commence your research (Saunders et al, 2003). The names of these labels or codes should ideally be derived from in vivo codes, which are the actual terms used by the participants (Strauss & Corbin, 1998).

Axial coding refers to the process of looking for relationships between the categories of data that have emerged from open coding, and it indicates a process of theoretical development as the fractured data are put back together in new ways by making connections between a category and its sub categories (Saunders et al, 2003). Sometimes as the coding proceeds, it will become clear that one of the codes needs to be partitioned into two or three subcategories, and, similarly, it may be that two codes that were originally separate are identified as related (Liampuntong & Ezzy, 2005). For example, in one's work project it may become apparent that the significance of a code for spending time at the gym is an aspect of self-care, because this activity involves physical activity to improve personal health.

The stage that follows is called selective coding, and this is intended to identify one of the principal categories, which becomes known as the central or core category, in order to relate the other categories to this with the intention of integrating the research and developing a grounded theory (Saunders et al, 2003).

It is cautioned that where you do not pay particular attention to the nature of the procedures outlined for grounded theory, you may not produce a research report that is sufficiently rigorous, to substantiate the explanation or theory that you are seeking to advance (Saunders et al, 2003).

Grounded theory analysis is intense and has set procedures to follow at each stage of analysis to arrive at a grounded explanation or theory. The stages of analysis include open coding, axial coding and selective coding.

It is acknowledged that as a distinct methodology, classic grounded theory is challenging to employ, and it can even be more challenging for a novice (Roderick, 2009). Moreover, there are different ways of doing grounded theory analysis (Liamputtong & Ezzy, 2005). This challenge increases further when graduate students encounter poor advice from dissertation supervisors who are unfamiliar with the methodology, or attempt to incorporate elements from the many alternative and modified versions of grounded theory presented in the literature (Roderick, 2009).

One came across an interesting article about the journey of a certain researcher through grounded theory. As she reflects on the course of her journey, Roderick (2009) observes that while there is a growing body of literature focusing on the experiences of learning to do qualitative research, little has been written about the experiences of learning classic grounded theory from the novice's perspective. Furthermore, graduate students who aspire to employ classic grounded theory for their theses or dissertations, predominantly learn the methodology on their own (Roderick, 2009). Few individuals have access to relevant graduate level courses or a dissertation supervisor experienced in classic grounded theory (Roderick, 2009). In fact, because of the many ways classic grounded theory has been altered and modified since Discovery of Grounded Theory was first published in 1967, many individuals who supervise classic grounded theory dissertations may have misunderstandings of the methodology (Roderick, 2009).

For example, a certain academic was once given a PhD thesis to read, of which his colleague assumed that he might be able to offer some useful insights since it was relevant to a study that this particular academic was engaged in (Deady, 2011). The methodology used by the PhD candidate was classic grounded theory, with which this particular academic had passing familiarity (Deady, 2011). Later on, this particular academic admitted that he had read the PhD candidate's thesis with the methodological bias of a phenomenological perspective and this meant that he had misinterpreted some commentary as subjective and wondered at the absence of other data (Deady, 2011). In short, this particular academic had misunderstood the methodology being used and so had missed the point of the argument being presented (Deady, 2011).

Some former graduate colleagues who employed qualitative approaches such as grounded theory, experienced more challenges and frustrations which were compounded by the difficulty to find an appropriate supervisor, as compared to those who employed positivist approaches.

Employing grounded theory can be very challenging and frustrating, and rather more so for a novice researcher. This challenge can be compounded if the novice researcher or graduate student is assigned a research supervisor who does not have a clear understanding of grounded theory or is biased towards a different epistemological approach.

It is acknowledged that data saturation or theoretical saturation is integral to naturalistic enquiry (Bowen, 2008).

Saturation occurs when additional information no longer generates new understanding, and at this point, it is time to wrap up the research (Liampuntong & Ezzy, 2005). This may happen early or it may occur after a while. Within focus groups, research projects that comprise more heterogeneous participants will require more total groups (Liampuntong & Ezzy, 2005). This is because the diversity in the group makes it more difficult to identify coherent sets of opinions. Across groups, projects that compare several distinct populations segments will need more total groups in order to achieve saturation within each segment (Liampuntong & Ezzy, 2005).

Once saturation has been reached, any additional participants, sessions or information will basically be a waste of time and energy. For example, in one of the qualitative studies conducted within the Southern Africa Development Community (SADC), it was realized at some point that the additional focus group sessions were not necessary, since they did not enrich the results, but merely prolonged the collection and analysis of the data. It is very possible that saturation had already been reached at that stage.

Interestingly, some researchers observe that the saturation concept remains nebulous and the process lacks systematization (Bowen, 2008). It is argued that saturation is mentioned in many qualitative research reports without any explanation of what it means and how it occurred, and this presents a challenge to qualitative researchers, especially in the absence of explicit guidelines for determining data or theoretical saturation (Bowen, 2008). While saturation has a distinct theoretically embedded meaning in grounded theory, its ubiquitous and non-selective use risks rendering the term meaningless to the qualitative research community (Caelli et al, 2003).

Saturation theory can be applied to many situations and is very important in qualitative research, because it entails gathering data to the point of diminishing returns, and also the point at which no new insights are obtained (Bowen, 2008). However, it is argued that the claims of saturation should be supported by an explanation of how saturation was achieved and substantiated by clear evidence of its occurrence (Bowen, 2008).

Although there are distinct strategies for analysing qualitative data, one particular feature that is common to these types of strategies involves the researcher in disaggregating the mass of qualitative data that she collects, as she collects them, into meaningful and related parts or categories (Saunders et al, 2003). This allows the researcher to rearrange and analyse these data systematically and rigorously (Saunders et al, 2003). Adopting this approach essentially means transforming the nature of the data that the researcher has collected, in order to allow the researcher to: comprehend and manage them; integrate related data drawn from different transcripts and notes; identify key themes or patterns from them for further exploration;

develop and/or test hypotheses based on these apparent patterns or relationships; draw and verify conclusions (Saunders et al, 2003).

There are many approaches, but one way is simply to use the research questions to group the data and then look for similarities and differences. This approach may be particularly appropriate when there is limited time or resources for a more in depth analysis or when your qualitative research is a smaller component of a larger quantitative study, conducted to provide further depth in predefined areas of interest (Ulin et al, 2005). For example, it would be interesting to see how the data would be analysed if a qualitative component was incorporated to the clinical trial of a vaginal microbicide product that has recently been conducted in South Africa.

A more in-depth and inductive approach to analysis is also appropriate, and this identifies a sequence of interrelated steps in data analysis: reading, coding, displaying, reducing, and interpreting. As you read and reread texts and review notes, you listen for emerging themes and begin to attach labels or codes to the chunks of text that represent those themes. Once your texts have been coded, you explore each thematic area, first displaying in detail for information relevant to each category and then reducing this information to its essential points (Ulin et al, 2005).

The categories that one devises need to be part of a coherent set so that they provide you with a well-structured, analytical framework to pursue your analysis. As your analysis develops you will develop a more hierarchical approach to the categorization of your data, whereby some category codes or labels will be developed and used to indicate analytical linkages between, and interpretations of, the data emerging (Saunders et al, 2003). This is what one would consider as the understanding of the individual components in the context of the bigger picture.

It is necessary to separate qualitative data into component parts so that those component parts can be understood individually and in the context of the bigger picture. After all, qualitative analysis emphasizes how data fit together as a whole, bringing together context and meaning, but it is cautioned that the use of language and constructs from the researcher's professional discourse to describe participants' meanings and actions should be avoided (Fossey et al, 2002).

After the researcher has compiled and processed all information, she can begin to code the themes. According to Ulin et al (2005), coding is the process of attaching labels to lines of text so that the researcher can group and compare similar or related pieces of information. Welman et al (2005) highlight that the following types of coding may be used: descriptive codes, interpretative codes, pattern codes, and revising codes. It may be a challenge to a novice researcher to reduce the huge amount of data to manageable and understandable texts. Welman et al (2005) observe that this challenge may be compounded by the fact that in qualitative analysis this is done chiefly with words, which may complicate the analysis because words usually have multiple meanings. Below are the coded transcripts of the three interviews that were conducted on 24 May 2022.

Site: Executive Office – Port of XX

Interview C001

Interviewer: YYZ Mabuz

Transcriber: YYZ Mabuz

Date: 24 May 2022

Start: 10h00

End: 10h23

Question 1

Interviewee: Can you please tell me what you understand about health	Coding
Respondent: I don't think I can give you a doctor's definition, but I can give you the definition of health from an operations manager's perspective. Let's see ... well, well, health depends on what you are referring to. Right now I would say if **you do not have any issues that bother you, then you are healthy, but if you are constantly worried, it means you are not healthy**. Does that make sense?	Health = no issues that bother or worry you
Interviewer: is that all you can tell me about your understanding of health?	Coding
Respondent: Yep! That's it. But perhaps one could add another dimension and say, if you guys declare us **fit for duty it means we are healthy**, but if you don't , then we are in trouble, because that would mean there is something-something somewhere not right. We have seen your monthly reports indicating that there is a growing trend of employees with **bp, diabetes, blah blah blah eehh** () that our employees are not attending the clinic as they should.	Health = fit for duty, without blood pressure, diabetes, etc.

Question 2

Interviewer: What would you regard as the pressing health issue in your company?	Coding
Respondent: Mmhh ... let me let me say ehh ... how can I put it? I know you guys are currently busy with the health awareness month sessions in the various departments of the company. I am sorry that I could not attend the sessions you had with the management team last Tuesday. You know, it is so hectic for me, cos	Burning health issue = no high disabling frequency rate

I've been attending incident investigation after incident investigation during the past two months due to the frequent number of disabling injuries on duty that we are faced with. You know what... I'm so frustrated. By the way, what was your question? Oh...yah, I remember. But I think I've already answered your question because judging from the high **disabling frequency rate in this company, it shows that we are not healthy**, all of us here. We are running like headless chickens. Eish... this thing frustrates me. Ok, let's carry on, as I have to rush to another meeting eeh... at half past

Interviewer: How does this affect the business? Coding

Respondent: The department of labour's eyes is on us. They are watching us very closely, which means that the company is in the black books. You can imagine. You can imagine for yourself, **our clients will not be too pleased to do business with us because we are now seen as an unsafe working environment**, especially for the past two months. Things have been really very bad for us. **Losing our clients will not be good because the business has been going through a bad patch** during the past twelve months. The volumes and tonnages are very low, and we were not able to reach our norms and stretch targets by far during the past financial year. This is so ironic because the business has been performing so brilliantly for so many years now, and there is already a huge budget set aside for further expansion of this port.

Impact of health on the business = creates good reputation, attracts clients

Question 3

Interviewer: If you regard this as the pressing health issue affecting the business, what has the company done to address this issue? Coding

Respondent: Obviously we have the **on-site clinic**, and we have approved that the **clinic operates 24 hours** per day to ensure that emergencies that happen at night are promptly attended to. What else? ()

Health intervention = 24 hours on-site clinic

Interviewer: How effective has this intervention been?

Respondent: Mmmhh... () how can I put it ... (). It is a bit tricky to tell because the clinic has been operating 24 hours for only three months now. But I could say we have not yet seen that much of an impact because we still have a **backlog of medicals** as you always report at the daily MDT.

Health intervention = 24 hours on-site clinic, but there is a backlog of medicals

Question 4

Interviewer: From your perspective, how would you ensure that there is an impact to ensure that the backlog of medicals is addressed promptly and that employees are fit for duty?

Coding

Respondent: You got me there... ehh... ehh... This a bit tricky. What can I say (). It means we have to stop the operations to allow people to visit the clinic. You know, **if we can release all the employees to go to the clinic at the same time, there will be huge delays in operations** as our clients also demand a quick turnaround. Time is of the essence here because any delay in operations will come back to haunt us as we will not be able to meet our norms and targets, and we may even incur huge penalties.

Fitness for duty = sending employees to the clinic would cause delays in operations

Interviewer: Ehhe ...

Respondent: If the business does not perform, my head is on the block. We are talking business here. There is a lot of pressure from clients and from head office. I do not want to be seen as a failure. At the end of the financial year, everybody expects to get a bonus, but they do not understand that it is through the hard work we put in operations () ... that makes the bonus possible.

Question 5

Interviewer: How integral is the health of employees to the success of the business or lack of?	Coding
Respondent: **That is why we have you and the clinic to ensure that you support us as a business** in that aspect.	Impact of health on the business = health is a support to the business, but responsibility shifted to health professionals
Interviewer: Please elaborate	
Respondent: I mean to say ... () **that is your competence area as health professionals to look after the health of our employees. That is why we employed you to look after that aspect**. You know! I am running late for the next meeting, please excuse me.	Impact of health on the business = responsibility of employee health shifted to health professionals, senior management

has no time
to discuss
health
matters

Interviewer: Thank you for your time Sir.

Respondent: Sharp Doc ... ().

Site: Corporate Affairs Office – Port of XX

Interview C002

Interviewer: YYZ Mabuz

Transcriber: YYZ Mabuz

Date: 24 May 2022

Start: 12h17

End: 12h53

Question 1

Interviewer: As the person in charge of communications in the business, what is your understanding of the health?

Coding

Respondent: To me health is a very broad term. I have learned a lot from doing my own research as you know that our portfolio requires us to ensure that we communicate correct information that has been verified to ensure that we do not send wrong messages that will create problems and confuse people, and to avoid tarnishing the name of the company. On the one hand, health means being at **peace with yourself and on the other hand health means being physically well, and ensuring that you eat well, you sleep well,** you have regular medical check-ups to ensure that there is **nothing wrong with you.** I guess it depends from which angle you look at it. There could be many definitions. But as far as I am concerned that is what I can describe health as.

Health = peace with yourself, physical wellbeing, eating well, sleeping well, nothing wrong with you

Interviewer: Okay ...

Coding

Respondent: Just one thing that comes to my mind ... () I remember last year, I thought I was healthy because I thought I was coping with the pressure at work. And all of a sudden, boom! I was hospitalized. It is only then that I realized that **I had burned both ends of my candle**. Anyway, I was still new in this job and was trying to prove a point that I could make an immediate impact.... ((Laughs loudly)). The hospitalization brought me down to reality. I wish we could all realise how vulnerable we are. I value my health so much now. It was such an unpleasant wake-up call when I got hospitalized, and that was my very first time ever to be

Health = not to burn both ends of the candle

admitted in hospital!

Question 2

Interviewer: What would you regard as the pressing health issue in your company right now?

Coding

Respondent: Mmhh (...) it's quite (..) interesting that you are asking me this question because it was highlighted at our EXCO meeting yesterday that **absenteeism is getting out of hand at this place**. Judging from the trends that were projected, I can conclude that our workforce is really sick. I have already put these absenteeism trends on the plasma screens and loo doors to make everybody in the business aware of the huge challenge that we face. Everybody must come on board and play a part.

Burning health issue = Absenteeism

Interviewer: Ehhe

Coding

Respondent: (...) the health communication partnership between the health department and our department has opened my eyes big time! I never imagined that **high blood pressure and diabetes** were so serious in this company. At least we are doing our best to communicate this message in all available platforms of the company. I am sure the change champions that have recently been launched will play a key role in changing the culture. It is evident that things have been taken for granted for too long in this company, specifically matters that touch on health (...) We as individuals as well, and I am testimony to that.

Burning health issue = high blood pressure, diabetes

Question 3

Interviewer: In your opinion, how integral is the health of employees to the performance of the business and why?

Coding

Respondent: It is very very integral because if you look at the high absenteeism trends that I have just mentioned, it means that the operations of the **business is compromised when workers are absent.** () I believe that **if workers are healthy and not frequently absent from work, the business can do very well.**	Impact of health to the business = business is compromised if workers are absent, business does well if workers are healthy and not absent
Interviewer: Please elaborate	Coding
Respondent: Sure (). The crux of the matter is that if you look at it on a broader scope, **if people are unhealthy whether they come to work or not the performance of the business gets affected** (...) Do you get my point? If you are sick and you are at work your performance is not 100 per cent. If your performance is not 100 per cent, it is obvious that the performance of the business will also not be 100 per cent. (...) obviously, when you are absent the business suffers because there is no one to perform your duties, even if they find a temporary replacement, that temporary replacement is likely to make mistakes. I have seen that even in my department whereby two of my workers were frequently absent last year, and this really hit us big time because I had to perform their duties as well as my duties, but I realized that some of the duties had to be compromised over others. I had to take a stand and recruit interns to understudy these frequently absent employees, to ensure that the work	Impact of health on the business = unhealthy workers affect performance of the business

carries on without much of a negative impact when one or both of my workers are absent. If we could feel the negative impact as a service department, how much more if the absenteeism happens in operations where the bread and butter of our business lies. It is much deeper, very much deeper (). Coming here has opened my eyes to things that I used to take for granted during my previous job as a freelance journalist. And (...). Okay okay

Question 4

Transcript	Coding
Interviewer: You have mentioned that the company is faced with the challenge of a high absenteeism rate due to the fact that the workforce is really sick. What has the company done to address this challenge of sick employees?	Coding
Respondent: The company has implemented the **return to work interviews for every employee who returns to work after having been off sick.** I have had a number of the return to work interviews with my two employees that were frequently off sick last year. What else has the company done (...) Yes, the company has also made it **compulsory for us as line managers to report on absenteeism on a weekly basis, and there have been absenteeism training for all line managers and supervisors.**	Health interventions = return to work interviews, compulsory weekly absenteeism reporting, training for managers and supervisors
Interviewer: Do you think these interventions have been effective, and why?	Coding
Respondent: To a large extent, I would say no because when you look at the absenteeism trends, the graphs	Health interventions

that we have projected on the plasma screens throughout the terminal, **the graphs show an increase in absenteeism instead of a decrease.** This shows that **our employees are getting sicker and sicker.**	= not effective, increase in absenteeism, employees getting more sick
Question 5	
Interviewer: What would you suggest or recommend as the most effective way of addressing the health issues affecting your company?	Coding
Respondent: Ye ye ye (...) it is a tough one (...) ((shaking her head)). Let me be controversial and **suggest that the company employs more women and younger people** as my general observation is that there are very very few women in this company and the majority of the current employees seem to be old. It was pretty obvious at the long service awards event held last week that we have a lot of workers who have worked for far too long here and their bodies are tired now. Where are the women? We need more of the women ((laughing loud))	Health interventions = recommend recruitment of more younger employees and women
Interviewer: Okay	Coding
Respondent: Honestly, I think that is an area that **needs to be looked into seriously.** Something **needs to be done very urgently**	Health intervention = needs serious and urgent attention

SITE: CORPORATE AFFAIRS Office – Port of XX

Interview C003

Interviewer: YYZ Mabuz

Transcriber: YYZ Mabuz

Date: 24 May 2022

Start: 14h04

End: 14h33

Question 1	Coding
Interviewer: As a full time representative of workers, what is your understanding of health Respondent: Thanks Doc, this is a profound and important question. Health is very close to my heart. My thinking is that health it is not easy to be completely healthy, and I am sure there is no single person in the world who can claim to be completely healthy. Having said that, I think it is important to distinguish things and compare apples to apples. There is a healthy person and there is a healthy environment. By healthy person I would say, that particular person has **no hindrances that affect his or her mind and her body**. A healthy environment is an environment where all people live and **work in harmony**, and there are **no hindrances of any kind** if we are to consider an environment to be healthy. In an environment such as this one at the port, I wouldn't regard it as a healthy environment because workers work under difficult conditions such as the dust, and harsh weather conditions. At the same time, the workers work very hard but the money is very little. Let me tell you a little story. (...) I remember when there was this long strike in 2010, when management had to wear overalls and did the manual work that the majority of our workers do on a daily basis, and they also felt how it is to be exposed to harsh weather conditions. Within a day, those managers could not take it any-more and some of them had to be admitted into hospital because they felt the strain of being exposed to unhealthy environmental conditions and of hard labour.	Health = body and mind Health = work in harmony, no hindrances

Interviewer: Okay	Coding
Respondent: It breaks my heart when I see our own African people who hold senior positions such as management disregarding the wellbeing of our fellow brothers and sisters, mothers and fathers, and fellow human beings in general. That is why our trade union is calling for the end of labour brokers because these labour brokers perpetuate the plight of our workers. In a nutshell, health to me means the **absence of any form of exploitation.**	Health = absence of exploitation
Question 2	
Interviewer: In your view, what would you regard as the most pressing health issue affecting this company that you work for?	Coding
Respondent: (...) **Exploitation of workers**, of course is a pressing health issue here. We may see other issues such as HIV and so forth and so forth, but the underlying factor is **what makes people more sick is the environment and the conditions in which they live and work**. I know that some people may dispute that, especially those that occupy management positions (...). But I maintain that point of view because it is the truth. The truth is painful, but I say it as it is. Our management knows deep down in their hearts that they are not in any way committed to making sure that the workers are healthy. Things are getting out of hand	Burning health issue = exploitation of workers, environmental conditions
Interviewer: Ehhe	Coding
Respondent: I must admit though that it would be unfair to shoulder all the blame on management, because we as workers also need to take responsibility	Burning health issue = lifestyle, not

of our own lives. **What we eat and how we behave** can be a factor on whether we feel healthy or not. At least being a change champion has helped me to look at things from a broader angle. Some people are just negligent of their own lives. They **do not want to check their health status**, and even if they have checked and been told that there is a problem they continue **living as if nothing was wrong.** I guess you have a huge load on your shoulders as it is not easy to change people.	checking health status
Question 3	
Interviewer: Exactly what has the company done to address health issues at the workplace?	Coding
Respondent: I must say that all workers are required to do **annual medicals at the company's clinic**. I am sure this has somewhat helped to pick up health conditions and to make people aware that they have such and such a condition. There are **education and awareness sessions done by the health team on a daily basis**. Right now the company has launched a hotline to assist workers and their family members who need assistance or advice on health issues and also a counseling service. You see, I am not just talking as a labour union representative, but am wearing both my hats as a labour union representative and as a change champion. That is fair, is it not? (...)	Health interventions = annual medicals at company clinic, education and awareness, hotline counseling service
Interviewer: Ehhe	Coding
Respondent: most of these interventions are still relatively new, and therefore it is not that easy to gauge their effectiveness at this stage. These things need the continuous full support of the powers that be.	Health interventions = change champions

Transcript	Coding
Otherwise, they just will fall flat. We are doing our bit as **change champions to support the cause**	
Question 4	
Interviewer: In your opinion, how integral is the health of employees to the performance of the business and why?	Coding
Respondent: **Employees are like the pillar of the company**. So, if the pillar is weak or if the employees are not healthy, the company will also be weak, but if the pillar is strong and the employees are healthy, then the performance of the business will also be strong. Yes, this shows how integral the health of employees is to the performance of the business.	Impact of health to the business = employees are the pillar
Interviewer: Okay	Coding
Respondent: (...) if a company does nothing to address the issue of health of its workers, such a company is doomed to failure. **Workers health is so central to any business performance**, no matter how small the company may be. In our case, the port is so huge, which means that health of workers is even more crucial.	Impact of health to the business = workers' health is central to business performance
Question 5	
Interviewer: In your opinion, what would be the best recommendation you can give to your company to ensure that all employees are fit for duty?	Coding
Respondent: (...) (...) ((scratching his head)) the best recommendation I can give is **regular check-ups at the clinic**. Oh! The other important thing is that the company must make it compulsory for all workers to **join medical aid**.	Fitness for duty = regular health check-up, joining medical aid

Interviewer: Okay	Coding
Respondent: The issue of medical aid is so crucial because if people do not belong to a medical aid they will have to face long queues in the state hospitals which do not often give a good service because they are so overcrowded. Right now, I went to visit one of our workers who is in a critical condition in one of the public hospitals and his family is so frustrated because the person has been lying there for almost three weeks and they were told that he is still waiting to do a scan. When we investigated, we discovered that the scan has been out of order since last year. It is such a crisis and drama in the state hospitals. So, I would emphasise that **every worker should be on medical aid**.	Fitness for duty = join medical aid
Interviewer: Thank you for your time Sir	Coding
Respondent: Eh (...) Thank you for the discussion, this issue makes me so emotional (...).	

According to Green & Thorogood (2009), indexing is a systematic way of applying codes or labels in the context of the overall data set. Below is an index of the data set derived from the coded transcripts in Part A above.

	Category/Themes	Subcategory/Subthemes
1.0	Meaning of health	
1.1		Peace with oneself
1.2		Physical and emotional wellbeing
1.3		Eating well
1.4		Sleeping well
1.5		No worries
2.0	Burning health issues	
2.1		Absenteeism
2.2		High blood pressure and diabetes
2.3		Unhealthy environment
3.0	Impact of health on business	
3.1		Healthy employees result in good business performance
3.2		Healthy employees result in attraction of new business and clients
4.0	Health interventions	
4.1		On-site clinic
4.2		Health education and awareness
4.3		Return to work interviews
4.4		Hotline counselling service
4.5		Change champions
5.0	Fitness for duty	
5.1		Regular health check-up
5.2		Joining medical aid

THE IMPORTANCE OF DATA management needs to be highlighted. This aspect of the project must be designed early so that systems are in place as soon as data collection begins (Cromar et al, 2006).

Ethics approval procedures with most authorities will require indication of how data will be handled and stored, and, how confidentiality will be safeguarded. Without a clear, ordered data-management system, data can easily be mislaid, or permanently lost, which could compound the inherent limitations of the qualitative data analysis approaches which could for instance result in the loss of some nuances of data such as in in thematic analysis.

When designing a data-management system it is important to think about how other members of the research or project team may be able to access the data in a clear and systematic manner (Cromar et al, 2006). As the analysis progresses, the integrity of the systems employed will become apparent, but it should be remembered that the data-management system used is a means to an end, not the end itself (Cromar et al, 2006). These systems may include index cards or computer-aided storage and retrieval systems (Liampuntong & Ezzy, 2005).

It is important that the data-management system stores original data, as well as research products or interim analysis, notes research journals, and analytical notes (Cromar et al, 2006). It is important that these documents be easily accessible to the other members of the research team and the researcher of course, but also kept as separate components of the research process (Cromar et al, 2006). However, it is also important to remember that the confidentiality of the information and the anonymity of the research participants must be protected at all times. Filing cabinets should always be locked and the databases password-protected (Cromar et al, 2006). Maintaining and securing such research documentation in a systematic fashion will result in a transparent research process and facilitate a more rigorous analytical approach (Ryan, 2009). There is now a range of computer assisted qualitative data analysis software available to assist in the process of qualitative data analysis (Ziebland & McPherson, 2006).

There may be a need to count the frequency of codes in thematic analysis. Computer software could come in handy at this stage, because it can easily perform statistical counts on the frequency of codes. However, novice researchers who are still new to qualitative research should be aware that these packages do not do the analysis, but merely assist in the analytical process through efficient storage and retrieval capabilities (Cromar et al, 2006).

Data management is crucial in qualitative analysis no matter which qualitative data analysis technique is employed.

References

Bowen, G.A. (2008) 'Naturalistic inquiry and the saturation concept: a research note', *Qualitative Research*, 8(1), pp. 137-152.

Caelli, K., Ray, L. & Mill, J. (2003) 'Clear as mud: toward greater clarity in generic quality research', *International Institute for Qualitative Methodology*, 2(2), pp. 1-13.

Cromar, N., Cameron, S. & Fallowfield, H. (2006) *Environmental health in Australia and New Zealand*. New York: Oxford.

Deady, R. (2011) 'Reading with methodological perspective bias: a journey into classic grounded theory', *The Grounded Theory Review*, 10(1), pp. 41-57.

Fossey, E., Harvey, C., McDermott, F. & Davidson, L. (2002) 'Understanding and evaluating qualitative research', *Australian and New Zealand Journal of Psychiatry*, 36(), pp. 717-732.

Green, T. & Thorogood, N. (2009) *Qualitative methods for health research*. London: Sage.

Jooste, K. (2009) *Leadership in health services management*. Cape Town: Juta

Joubert, G. & Erlich, R. (2007) *Epidemiology: a research manual for South Africa*. New York: Oxford.

La Pelle, N. (2004) 'Simplifying qualitative data analysis using general purpose software tools', *Field Methods*, 16(1), pp. 85-108.

Liampuntong, P. & Ezzy, D. (2005) *Qualitative research methods*. New York: Oxford.

Pope, C., Ziebland, S. & Mays, N. (2000) 'Qualitative research in health care: analyzing qualitative data', *BMJ*, 320(7227), pp. 114-116.

Roderick, C. (2009) 'Learning classic grounded theory: an account of the journey and advice for new researchers', *The Grounded Theory Review*, 8(2), pp. 49-63.

Ryan, M. (2009) 'Making visible the coding process: using qualitative data software in a post-structural study', *Issues in Educational Research*, 19(2), pp. 142-161.

Saunders, M., Lewis, P. & Thornhill, A. (2003) *Research methods for business students*. Harlow: Prentice Hall.

Strauss, A. & Corbin, J. (1998) *Basics of qualitative research*. Thousand Oaks: Sage.

Ulin, P.R., Robinson, E.T. & Tolley, F.E. (2005) *Qualitative methods in public health*. San Francisco: Wiley.

Welman, C., Kruger, P. & Mitchell, B. (2005) *Research methodology*. Cape Town: Oxford.

Ziebland, S. & McPherson, A. (2006) 'Making sense of qualitative data analysis: an introduction with illustrations from DIPEx (personal experiences of health and illness)', *Medical Education*, 40, pp. 405-414.

Chapter 6

Validity and Presentation of Findings to Different Audiences

The topics of discussion in his chapter are:

- Approaches to appraise qualitative research
- Validity and reliability in qualitative research

6.1 Approaches to appraise qualitative research

LITVIA & JACOBY (2002) suggest that credibility, dependability, confirmability and transferability are the approaches that should be used to appraise qualitative research, and that thick description, triangulation, reflexivity and member-checking can be the techniques used during the research process.

Credibility, also known as truth value, determines how the findings and conclusions of the study stand the closest scrutiny, and this can be seen as a corresponding criterion to validity in quantitative research (Litva & Jacoby, 2002).

It is acknowledged that qualitative and quantitative criteria of excellence are equally important but inherently different. As such, it is believed that the quality of qualitative research should be judged by the trustworthiness of the findings (Ulin et al, 2005). Even though the aspects of trustworthiness are separate, they should be viewed as intertwined and interrelated. Transferability is one of the approaches used to assess the true value or the trustworthiness of qualitative findings (Litva & Jacoby, 2002). Transferability is considered as the qualitative analogue of the concept of generalizability, and it is to do with how the findings of the study can be applied to other contexts (Litva & Jacoby, 2002). According to Graneheim & Lundman (2004), transferability refers to the extent which the findings can be transferred to other settings or groups. The authors can give suggestions about transferability, but it is the reader's decision whether or not the findings are transferable to another context (Graneheim & Lundman, 2004).

The importance of context in qualitative studies leads some researchers to doubt that results from one context should be transferred to another, while it may lead others to apply conclusions from their data too casually (Ulin et al, 2005). Because context is a key influence in any qualitative research, the researcher must account for contextual factors when transferring data from one situation to another (Ulin et al, 2005). Repeating the study in another population, with similar conclusions, lends credibility to the results and further specifies the circumstances under which the findings will occur (Ulin et al, 2005). Thus, well documented knowledge might be extended to similar populations, but the burden of proof lies with the original investigator than with the person seeking to make application elsewhere. The original enquirer cannot know the sites to which transferability might be sought, but appliers can (Ulin et al, 2005).

To facilitate transferability, it is valuable to give a clear and distinct description of culture and context selection and characteristics of participants, data collection and process of analysis (Graneheim & Lundman, 2004). A rich and vigorous presentation of the findings together with appropriate quotations will also enhance transferability (Graneheim & Lundman, 2004). Other authors add that the strategy of thick descriptions is essential in the context of the concept of transferability (Houghton et al, 2013).

One would rather take middle ground and say, lessons learnt from qualitative studies can be applied to other contexts if samples have been carefully selected to represent viewpoints and experiences that reflect key issues in the research problem. Ultimately, the goal would be to produce data that are conceptually, not statistically, representative of people in a specific context.

Transferability is part of the standards for evaluating the rigour of qualitative studies that are consistent with the worldview, information sources, and methods of the interpretivist paradigm. Although transferability may be regarded as a qualitative analogue to the concept of generalizability, generalizability is not relevant to the goals of the methodology of most interpretive work, it is nevertheless important to know whether the conclusions of a study are transferable to other contexts (Miles & Huberman, 1994).

Dependability is considered as the qualitative analogue to the concept of reliability, and it is to do with whether the research process was consistently carried out and in accordance with the rules and conventions of qualitative research given the instability and idiosyncrasies of the social world (Litva & Jacoby, 2002).

Confirmability refers to how the researcher has influenced the research findings (Litva & Jacoby, 2002). Confirmability thus means a way of knowing that, even as a co-participant in the inquiry, the researcher has maintained the distinction between personal values and those of the study participants (Ulin et al, 2005).

Thick description is about the detail of every step of the research process, and this includes the research context, site selection, research strategy, assessment of informants, data collection and recording, and data analysis (Litva & Jacoby, 2002). Thick-description is heavily used to enhance transferability, and it may also be used to enhance confirmability.

Triangulation is considered as an important technique to enhance credibility of the qualitative study, and this is particularly so because triangulation uses different methods, different researchers, different sources, and theories, to compare or confirm findings (Litvia & Jacoby, 2002). In this context, the use of different methods is known as method triangulation; the use of different researchers or investigators is known as investigator/researcher triangulation; the use of different sources is known as source/data triangulation; the use of different perspectives or theoretical points of view is known as theoretical triangulation (Litvia & Jacoby, 2002). Researcher triangulation can also be employed to enhance the dependability of qualitative research (Litvia & Jacoby, 2002). However, other authors argue that it is not practical to conduct triangulation because data come in different forms and defy direct comparison (Barbour, 2001).

Reflexivity refers to the self-consciousness of the researcher as she reflects about whether her assumptions have shaped the research output, bearing in mind that she was also part and parcel of the research (Green & Thorogood, 2009). Reflexivity can be used to enhance confirmability (Litvia & Jacoby, 2002).

Reflexivity has indeed become an important concept in qualitative research. Reflexivity – the researcher's critical self-awareness – is a vital process in which you as the researcher question and observe yourself at the same time you listen and observe the participant (Ulin et al, 2005). With its emphasis on egalitarian relationships, feminist methodology has contributed greatly to this point (Ulin et al, 2005). In contrast to the detachment required in many quantitative studies, your presence is a vital component of the qualitative research process in two ways (Ulin et al, 2005).

First, the researcher is in a partnership with the participant – working together to explore themes and find answers (Ulin et al, 2005). This is what Green & Thorogood (2009) regard as the first stage of reflexivity, which involves reflecting critically on the research itself. In this regard, the reflective researcher considers the broader politico-social context of their research, and may consider questions such as why particular questions are legitimate and why they can attract funding and interest in the findings (Green & Thorogood, 2009).

It is generally a challenge to get funding for qualitative research, so, therefore, if qualitative researchers could be more reflective about the research itself from a broader politico-social context, perhaps, this could enhance the chances of getting funding for future qualitative studies. If there could be unity in qualitative research, and such reflexivity enhanced, this could possibly result in qualitative research being seen in a more positive light.

Second, as you listen to your participant-partner, interpret, and respond, you are a key research instrument yourself, not only absorbing information but also influencing how it is elicited (Ulin et al, 2009). This is what Green & Thorogood (2009) regard as the second level of reflexivity which is more personal, and indicates that who you are and where you are as researcher will inevitably shape the kind of data generated. After all, this explicitly accounts for the fact that qualitative data is not merely collected but generated (Green & Thorogood, 2009). In being explicit, the researcher is attempting to ensure that the findings are reflective of the participants', rather than the researcher's perspective.

In essence, the researcher has to be up-front about the reflexive process to demonstrate her learning or change in thinking that has come about through engagement with participants in the study (Curtin & Fossey, 2007). Self-examination, documented with other observations in the field notes, is part of the iterative process of interpretation and revision that moves the data collection toward its goal (Ulin et al, 2005).

Reflexivity has indeed become an important concept in qualitative research as it enables the researcher to reflect on what impact she has made on shaping the research output (Litva & Jacoby, 2002). There is no doubt that reflexivity is one of the important techniques that can be carried out during the qualitative research process.

Member-checking is the process of feeding back the researcher's interpretations to the informants so that they can confirm or not confirm it (Litvia & Jacoby, 2002). Member-checking can be used to enhance dependability (Litvia & Jacoby, 2002). It may not be practical to conduct member checking because this could be time-consuming, or the informants may no longer be interested in the research (Litvia & Jacoby, 2002).

One does not think any of the approaches or techniques is more important than the other because each approach is important in the context of the type of qualitative research, the setting, and other unique factors.

6.2 Validity and reliability in qualitative research

THE ARGUMENT ON BIAS in contemporary qualitative literature has largely centred on bias during the research process, and the issue of bias that the researcher brings to a study when choosing a research topic (Deady, 2011). According to Kisely & Kendall (2011), a qualitative researcher accepts that there are multiple realities, not just one objective reality, and in particular, the truth is in the informants' perspective, not that of the assessors.

It is interesting that some researchers observe that some participants or interviewees may not tell the truth unless some level of trust has been established (Liamputtong & Ezzy, 2005). Is that perhaps why the participant's views are sometimes not held as the final validating authority? People are generally uncomfortable to share their personal information with strangers.

At the heart of the problem with the concepts of validity and reliability in qualitative research is the relationship between the observer and observed reality (Liamputtong & Ezzy, 2005). Understandings and interpretations change, and may be different in different social locations and may even differ from time to time for the same person in the same social situation (Liamputtong & Ezzy, 2005). This signals the importance of ensuring trustworthiness of the research findings.

In qualitative research there are basically four criteria - credibility, dependability, confirmability and transferability - that can be incorporated into the analysis in order to assess the truth value of qualitative findings. There are several techniques for ensuring that interpretation is credible: looking for negative cases for emerging hypothesis; testing rival explanations; and seeking explanations for inconsistencies arising from triangulation of respondents, methods, theories, or researchers (Ulin et al, 2005).

To increase dependability of qualitative findings, you might incorporate a team approach or use multiple independent coders or analysts (Ulin et al, 2005). This tactic will help to offset the subjective bias of any one researcher (Ulin et al, 2005).

Qualitative researchers do not claim to be detached and neutral scientists, unencumbered by their own experiences and values (Ulin et al, 2005). They do believe, however, that by being conscious of their own subjectivity, they can better understand and limit its effects on their research activities (from data collection to analysis), thereby allowing participants to express their experiences, values, and expectations without constraint (Ulin et al, 2005). Qualitative researchers can check whether they have sufficiently maintained the distinction between their own and their subjects' ideas by opening the study process to outside inspection and verification (Ulin et al, 2005).

Qualitative analysis can be intense and may seem confusing at times because of the element of subjectivity. Whether it is achievable or desirable to attain a state of complete non-bias is at best questionable and remains an ongoing debate (Deady, 2011).

An important test of quantitative reliability is the extent to which findings can be replicated. For qualitative researchers inquiring into unique constellations of multiple phenomena and meanings, this goal would be meaningless. For qualitative researchers the methodological parallel to reliability is whether the results are dependable, whether the research process is consistent and carried out with careful attention to the rules and conventions of qualitative methodology (Ulin et al, 2005). We ask ourselves: Are the research questions clear and logically connected to the research purpose and design? Are there parallels across data sources? Do multiple field—workers have comparable data collection protocols? (Ulin et al, 2005).

Saunders et al (2003) add that one response to make to the issue of reliability is that the findings derived from using non-standardised research methods are not necessary intended to be repeatable since they reflect reality at the time they were collected, in a situation that may be subject to change (Saunders et al, 2003). The assumption behind this type of research is that the circumstances to be explored are complex and dynamic (Saunders et al, 2003). The value of using this non-standardised approach is derived from the flexibility that you may use to explore the complexity of the topic (Saunders et al, 2003). This seems to affirm the point about dependability being assured through flexibility in the design. Therefore, an attempt to ensure that qualitative, non-standardised research could be replicated by other researchers would not be realistic or feasible without undermining the strength of this type of research (Saunders et al, 2003). It is suggested that researchers using a qualitative, non-standardised approach need to make this clear – perhaps, to transform an aspect perceived to be a weakness by some into a strength based on realistic assumptions about the ability to replicate research findings (Saunders et al, 2003).

A colleague's observation was interesting when she asserted that dependability can be assessed based on the auditability of the results, researcher triangulation, thick descriptions, and other strategies. However, one must question the feasibility of employing such a costly process, both in terms of time and money, to verify the reliability of a qualitative study (Carr, 1994).

In a nutshell, dependability may enhance rigour in qualitative research, but this also poses challenges. The reliability of qualitative research is weakened by the fact that the process is under-standardised and relies on the insights and the abilities of the observer, thus making an assessment of reliability difficult.

A question may be asked as to whether objectivity is affected if the background of the researcher is similar to that of the participant refers. The issue is not about objectivity per se, as qualitative research does not profess to be objective. Qualitative research such as participant observation entails subjective interpretation and therefore the potential for bias, but the researcher's subjectivities, including values, beliefs, and emotions should be accepted as centrally involved in the research process, and the notion of objectivity rejected as neither necessary, nor even desirable (Hewitt, 2007).

The issue is therefore about enhancing the rigour and integrity of a qualitative researcher such as in a participant observer study. According to Ulin et al (2005), one of the strategies to enhance the rigor of a participant observer study is to ensure that there is careful documentation of the observations. There is a possibility that a researcher who shares a similar background as the participants may take things for granted and not record everything that she observes, thinking that she would remember it when she compiles the report later. By neglecting to take this important step, the researcher risks losing what she may have learned or keeping valuable information in her mind, unavailable to the rest of the team (Ulin et al, 2005). Lost data can have serious consequences in misinterpretation of other data and lead to erroneous or incomplete conclusions (Ulin et al, 2005). This also leaves the researcher vulnerable to claims that her research may not be reliable. This makes me realise the magnitude of the challenge faced by qualitative researchers especially if their research is viewed from the perspective of the dominant positivist paradigm which subscribes to objectivity in research. A few years ago, one attended a seminar in Durban, South Africa, and was surprised to hear one of the speakers discouraging those who are aspiring to pursue qualitative research because he said it was not easy to get funding for such research.

The solution would be to create an audit trail, where the researcher would document her observations and conclusions in such a way that other researchers will be able to reconstruct the process that data that has led her to the results (Ulin et al, 2005). Other authors suggest that repeated conversations with the same participants, member checking and validation of data between researcher and participants, and using multiple sources of data could enhance the credibility and integrity of the qualitative study such as in participant observation (Averill, 2002).

Whether or not the researcher shares a similar background with the participants, objectivity in qualitative research is not really an issue as qualitative research is inherently subjective, but the issue is to enhance the research rigor and integrity of the qualitative researcher such as in participant observation.

References

Averill, J.B. (2002) 'Voices from the Gila: a health care issues for rural elders in south-western New Mexico', *Journal of Advanced Nursing*, 40(6), pp. 654-662.

Barbour, R. (2001) 'Checklists for improving rigour in qualitative research: a case of the tail wagging the dog', *BMJ*, 322, pp. 1115-1117.

Carr, T. (1994) 'The strengths and weaknesses of quantitative and qualitative research: what method for nursing', *Journal of Advanced Nursing*, 20, pp. 716-721.

Curtin, M. & Fossey, E. (2007) 'Appraising the trustworthiness of qualitative studies: guidelines for occupational therapists', *Australian Occupational Therapy Journal*, 54, pp. 88-94.

Deady, R. (2011) 'Reading with methodological perspective bias: a journey into classic grounded theory', *The Grounded Theory Review*, 10(1), pp. 41-57.

Graneheim, U.H. & Lundman, B. (2004) 'Qualitative content analysis in nursing research: concepts, procedures and measures to achieve trustworthiness', *Nurse Education Today*, 24, pp. 105-112.

Green, T. & Thorogood, N. (2009) *Qualitative methods for health research*. London: Sage.

Hewitt, J. (2007) 'Ethical components of researcher-researched relationships in qualitative interviewing', *Qualitative Health Research*, 17(8), pp. 1149-1159.

Houghton, C., Casey, D., Shaw, D. & Murphy, K. (2013) 'Rigour in qualitative case-study research', *Nurse Researcher*, 20(4), pp. 12-17.

Kisley, S. & Kendall, E. (2011) 'Critically appraising qualitative research: a guide for clinicians more familiar with quantitative techniques', *Australasian Psychiatry*, 19(4), 364-367.

Liamputtong, P. & Ezzy, D. (2005) *Qualitative research methods*. New York: Oxford.

Litvia, A. & Jacoby, A. (2002) 'Qualitative methods and evidence-based practice'. In: Craig, J. & Smyth, R. *The evidence based practice manual for nurses*. Edinburgh: Elsevier Science, pp. 136-163.

Miles, M.B. & Huberman, B. (1994) Qualitative data analysis: an expanded sourcebook. Thousand Oaks: Sage.

Saunders, M., Lewis, P. & Thornhill, A. (2003) *Research methods for business students*. Harlow: Prentice Hall.

Ulin, P.R., Robinson, E.T. & Tolley, E.E. (2005) *Qualitative methods in public health*. San Francisco: Wiley

PART 2: QUANTITATIVE METHODS IN PUBLIC HEALTH RESEARCH

Chapter 7

Scientific Method and Introductory Concepts

This chapter will cover the following topics:

- Approaches to scientific research
- Epidemiology, incidence, prevalence, and the concepts of prevention

7.1 Approaches to scientific research

IT HAS ALWAYS MADE one wonder why certain schools of thought are more dominant and more acceptable than others when it comes to research approaches. Bruce, Pope & Stanistreet (2008) acknowledge that paradigm shifts, social influences and experiences have had a huge bearing on how research has been conducted over the years. Although the status quo indicates that the dominant fields of biomedicine and epidemiology follow a positivist approach while the less dominant social sciences follow an interpretative approach, there is an emerging discourse which promotes a combination of research approaches (Johnson, Onwuegbuzie & Turner, 2007). As such, one's research experiences/interests to date might influence the way in which one decides to design and carry out future research.

Suppose one's career started in the biomedical sciences, and the only research approach used was a positivist approach, whereby all our research projects were conducted by following quantitative research methods. For example, if one was involved with molecular research and also spent some time in the cancer research laboratory whereby a series of laboratory analyses on glutathione-dependent metabolism of certain anticancer drugs were conducted. In such cases, quantitative research methods were the most ideal because there was an obligation to measure and to make calculations based on the molecular systems that we were dealing with.

After having qualified as a bio-chemical scientist and physician, and having gained clinical and community experience, one realised that human beings are more complex and unique. Although one still saw the important role of biomedical science and 'mainstream' medicine, one also felt that this was not enough because there was still much more that we did not understand about the uniqueness and complexity of human beings and about what makes them healthy or unhealthy. Human beings respond differently to similar stressors, diseases, disasters, and treatment. Hence, the need to follow an interpretative research approach that could explore and tap into the uniqueness of human beings and about what makes them healthy or unhealthy, without having to follow a formalized and standardized quantitative approach. One has subsequently completed a few qualitative research studies at different firms and communities of southern Africa, in order to explore the perceptions of people about the effectiveness of organisational health and wellbeing, social and economic empowerment, health strategies and programmes, among others. Lesson learnt was that not everything about human beings can be measured or generalized, especially when it comes to issues of health. Da Costa & French (1990) report that formalized models are not everything when it comes to health research. Sachs & Kranz (1997) highlight the importance of anthropology and sociology when studying a medical problem.

Certainly, one research approach cannot provide all the answers. As such, both an epidemiological and a socio-economic research perspective will be of value not only to the local community, but to the country, the region and beyond, in terms of interventions. A paradigm shift is necessary if we are to make a significant contribution to health, but this will require the methodological wrangles between the positivist and interpretative proponents to be minimized, in order to create an enabling environment for harmonisation, innovation and positive impact for health (Popay & Williams, 1996).

It is a pity that purists often believe that their way is the best way or the only way, and that their paradigm cannot coexist with another. There is no doubt that the positivist paradigm in research has enjoyed great dominance over the years and this has entrenched the agenda of quantitative research methods with the belief that research must be limited to what can be observed and measured objectively (Joubert & Ehrlich,2007) . It has been highlighted that Thomas Kuhn (1922-1996) popularized the idea of a paradigm and mentioned that any dominant paradigm is bound to be challenged and replaced by another at some point in time (Bruce et al, 2008). It is hardly surprising that the positivist paradigm has been challenged by social scientists and other interpretative proponents who argue that human behaviour cannot be simply reduced to a mere value or statistic (Welman et al, 2005).

Human beings need to be understood not only from a positivist perspective, but from a holistic perspective, and a mixed research approach could be of benefit in that regard. Philosophers such as Charles Sanders Peirce and John Dewey advocate a pragmatic stance in research (Johnson & Onwuegbuzie 2004). The weaknesses of one paradigm could be complemented by the strengths of the other paradigm. For example, during one's time as a field researcher to determine the effectiveness of family planning methods in the rural, urban and industrial settings of the Kingdom of Eswatini, one realized that both approaches nicely complemented each other. In addition to making quantitative analyses, we were able to qualitatively gain an insight into why people behaved the way they did according to their own experiences in their own settings. This helped in understanding local factors such as human experiences and other factors, for intervention purposes (Sachs & Kranz, 1997).

Moreover, both research paradigms are important, and the unique strengths of each paradigm could benefit the overall human behaviour research discourse. It is interesting to observe that there are many researchers who have now realized the importance of having an integrated research approach. Johnson & Onwuegbuzie (2004) observe that contrary to purists' beliefs, the two paradigms have important similarities such as empirical observations and use of safeguards, which are an opportunity to strengthen the new discourse of a mixed method research approach.

It is imperative that proponents of the two research paradigms get their house in order for a common purpose as we are now faced with new situations in public health research and practice (Popay & Williams, 1996). We need to be open-minded and transcend the disciplinary boundaries and realize that there are other ways of conducting research, which could benefit the overall health research and thereby balance the communication and understanding of human behaviour research (Rosenfield, 1992). However, it has to be acknowledged that human behaviour cannot be completely understood as human beings are unique and influenced by complex factors.

It is a scary reality that the force of gravity is pulling us in one direction, as we have to make do with what we have at our disposal, and what we think is the best science of our time. The good thing is that science is not fixed as it is always evolving, and as it evolves we also evolve and hopefully gain a better understanding. At the same time, one wonders if what we consider as the best science of our time is indeed the best science of our time. Is there a possibility that most published health research findings will be proved wrong in the future as the paradigm of scientific research evolves? One will probably say yes, because one believes that nothing is absolute in this world.

There are many things that cannot be measured using the current scientific methods. In fact, there is a lot that we still do not understand about life. Some scholars acknowledge that epidemiology does not seek to find proof, but seeks to find associations, even though epidemiologists put more emphasis on empiricism. However, there are times when the epidemiologist takes things for granted and believes that everything must be evidence based (Bruce et al, 2008). Similarly, there are times when those who follow a qualitative scientific approach also take things for granted and conduct their research with a perceived agenda with regard to the research question instead of first getting to understand the needs of the target population (Maghboeba et al, 2005).

It gets fascinating at times to observe the power of the media in driving the agenda of evidence-based medicine or science which derives from the scientific notion that everything must be objectively measured and tested to prove its efficacy. The issue of evidence-base has had a huge bearing on research funding. Anything that is portrayed to be outside the scope of the evidence base is often side-lined. Money can affect the norms and the philosophy of science especially with the involvement of the multi-billion pharmaceutical industry in health care research and funding. Political lobbying can also influence the dimension and direction of the philosophy of science as this can influence the national or international health budget which can sway the power in favour of what is considered an acceptable scientific paradigm of that particular time.

When one reflects on the work of French philosophers such as Deleuze and Guattari, one realizes that health sciences are outrageously exclusionary and dangerously normative when it comes to scientific knowledge and scientific research (Holmes et al, 2006). It was interesting to read a few articles reporting that the majority of published research results are false, but they are accurate based on the prevailing scientific bias or paradigm (Ioannidis, 2005).

7.2 Epidemiology, Incidence, prevalence, and the concepts of prevention

THERE ARE MANY GOOD definitions of epidemiology that have been proposed. However, the one that captures the spirit in the context of public health research is: 'epidemiology is the study of the distribution and determinants of health-related states or events in specific populations, and the application of this study to the control of health problems' (Dictionary of epidemiology, 2001). As such, epidemiology is not about individuals, but it is a population science.

Statistics is a discipline which includes empirical and estimated quantitative data, and the classification of such data according to probability theory and hypothesis testing (World Health Organization, 2018). For example, health statistics includes empirical data and estimates related to health, such as mortality, morbidity, risk factors, among others, to display, summarise and compare data, and makes it possible to make inferences about a population or populations (World Health Organization, 2018). Experts in statistics sometimes regard statistics about births, mortality, marriages and divorces as vital statistics (U.S. National Library of Medicine, 2017).

Epidemiology and statistics are related in the sense that epidemiology is a quantitative discipline which relies on statistical approaches or models to analyse data, applicable to sound research methods.

Scenarios of incidence and prevalence

- The term incidence in relation to angina could be explained as the number of new cases of angina in a defined at risk population in a specified period of time; whereas, prevalence is all the cases of angina that existed in that defined at risk population at that particular point in time or during that specified period (Bruce et al, 2008).

- This might be because prevalence includes all the cases of angina that existed in 2009 including cases of angina that existed at the start of 2009 and new cases that occurred during 2009. Joubert & Ehrlich (2007) refer to this type of prevalence as period prevalence. On the other hand, the incidence of angina excludes cases of angina that existed at the start of 2009 and only included new cases of angina that occurred during the course of 2009. Unlike prevalence, incidence often focuses on a smaller population (Waraich et al, 2004). Nevertheless, both prevalence and incidence are useful ways of measuring disease frequency (Gavin et al, 2005).

- Calculation of the prevalence of psychiatric disorder in residential care:

Prevalence = Number of cases at a given time/Number in population at that time (Bruce et al, 2008)

Given number of cases at a given time = 198

Given number in population = 2,790

This implies that prevalence = 198/2,790 = 0.07

Therefore, the prevalence of psychiatric disorder amongst those in residential care is 7 per cent or 70 per 1000

- Calculation of the prevalence of psychiatric disorder in foster care:

Prevalence = Number of cases at a given time/Number in population at that time (Bruce et al, 2008)

Given number of cases at a given time = 324

Given number in population = 8,390

This implies that prevalence = 324/8,390 = 0.039

Therefore, the prevalence of psychiatric disorder amongst those in foster care is 3.9 per cent or 39 per 1000

Scenarios of the concepts of prevention

1. Offering alendronate (a drug to prevent osteoporotic fragility fractures) to women with osteoporosis to prevent fracture	Tertiary prevention
1. Screening using mammography to identify the presence of early breast cancer and subsequent treatment of positive cases	Secondary prevention
1. Approaches such as smoking prevention and cessation treatments in a group of asymptomatic smokers to reduce the risk of lung cancer	Primary prevention
1. In a group of patients suffering from incontinence, carrying out bladder training exercises to prevent the condition from worsening	Tertiary prevention
1. The application of dental sealants in school children to prevent tooth decay	Primary prevention
1. Faecal occult blood testing to screen for colon cancer to reduce the incidence and mortality of colon cancer through the detection and removal of precancerous polyps	Secondary prevention

1. Early detection and treatment of diabetic eye disease to reduce the possibility of blindness in patients with diabetes	Tertiary prevention
1. Palliative care in patients with bowel cancer to reduce inflammation (corticosteroids) and pain (morphine)	Tertiary prevention
1. Establishing an antibacterial water treatment intervention to provide safe drinking water in a village in Guatemala and reduce the occurrence of diarrhoea in children in the village	Primary prevention
1. Prescription of anti-inflammatory drugs in patients with rheumatoid arthritis to improve range of movement and reduce pain	Tertiary prevention

It is possible that the discovery and provision of antiretroviral drugs (ARVs) as treatment for HIV has been both a blessing and a curse.

The blessing has been significant as ARVs have kept people who are living with HIV, healthy, and reduced the chance of passing HIV on to others. As such, ARVs have given hope to individuals, families, communities, workplaces, nations, and the world at large, as huge reductions in transmission, morbidity and mortality rates have been seen where antiretroviral therapy (ART) has been used correctly and consistently.

The curse has manifested in a number of ways, particularly in the heavily disease-burdened southern Africa region, and other disease burdened regions in different parts of the world. Firstly, knowing that many people who take ARVs as treatment can live for many years, and that those who take ARVs as prevention (such as pre-exposure prophylaxis - PrEP, and/or post-exposure prophylaxis - PEP) could result in people reverting to risky sexual behaviour or risky drug injecting behaviour with the comfort that there are ARVs available. Such risky behaviour could become fertile ground for further spread of HIV. Secondly, the ARVs are very expensive, and most of the countries south of the Sahara are poor and cannot sustain the supply of ARVs, and as such, lack of sustainable supply could compound the challenge of poor adherence to treatment. Thirdly, a huge chunk of the health budget of most southern African countries is channelled towards HIV and AIDS, and this somewhat compromises other key strategic and programmatic areas that need urgent attention such as the epidemic of non-communicable diseases and emerging/or re-emerging infectious diseases (Sewankambo & Katamba, 2009).

In 2014, the news that the "Mississippi Baby" who was put on antiretroviral treatment within 30 hours of birth and was supposedly "cured" of HIV, was met with great excitement and celebration by the majority of people worldwide. Similarly, there was great excitement and belief that scientists were very close to finding a cure for HIV when it was reported at the 9th International IAS Conference on HIV Science, in Paris, 23-26 July 2017, that a nine-year old South African child with HIV at birth and who had been off antiretroviral drugs for eight and a half years showed no symptoms or signs of active HIV. To see such great joy in the faces of many people was really very interesting because it made one realize the magnitude and the impact those positive stories had made. If scientific research could come up with a cure for HIV, who will really benefit? Some studies report that it is the scientific innovators or the drug producers that would benefit the most because there will always be a huge demand particularly from the middle- and low-income countries where there is the heaviest burden of HIV and AIDS (Philipson & Jena, 2005).

When we talk about a cure, it makes one think, because we seem to live in a society which believes that everything must have a cure. The concern is that in due course, the cure is bound to fail or result in therapeutic drug resistance, which means, scientific research must again look for another cure to deal with an even tougher "monster" that has been created. For example, it is a known fact that mycobacterium tuberculosis (TB) can be cured. However, there are already tougher "monsters" such as multi-drug resistant tuberculosis (MDR-TB) and extreme drug resistant tuberculosis (XDR-TB) cases that we have already seen. Is a cure for HIV not going to create another tougher "monster"? Only time will tell. The bottom line is that prevention is better than cure. Primary prevention is the most preferred because it prevents a disease from the beginning, then secondary prevention, and lastly, tertiary prevention (Bruce et al, 2008).

References

Bruce, N., Pope, D. & Stanistreet, D. (2008) Philosophy of science and introduction to epidemiology'. In: *Quantitative research methods for health research: a practical guide to epidemiology*. Chichester: John Wiley & Sons, Ltd.

Da Costa, N.C.A. & French, S. (1990) 'The model-theoretic approach in the philosophy of Science', *Philosophy of Science*, 57 (2), pp. 248-265.

Gavin, N., Gaynes, B., Lohr, K., Meltzer-Brody, S., Gartlehmer, G. & Swinson, T. (2005) 'Perinatal depression: a systematic review of prevalence and incidence', *Obstetrics & Gynaecology*, 106 (5), pp. 1071-1083.

Holmes, D., Murray, S.J., Perron, A. & Rail, G. (2006) 'Deconstructing the evidence-based discourse in health sciences: truth, power and fascism', *International Journal of Evidence-Based Healthcare*, 4 (3), pp. 180-186.

Ioannidis, J.P.A. (2005) 'Why most published research findings are false', *PLoS Medicine*, 2 (8), pp. 124.

Johnson, R.B., Onwuegbuzie, A.J. & Turner, L.A. (2007) 'Toward a definition of mixed methods research', *Journal of Mixed Methods Research*, 1 (2), pp. 112-133.

Joubert, G. & Ehrlich, R. (2007) *Epidemiology: a research manual for South Africa*. 2nd ed. Cape Town: Oxford University Press Southern Africa.

Last, J.M. (2001) Dictionary of epidemiology. 4th ed. New York: Oxford University Press, p.61.

Maghboeba, M., Simon, C., Van Stade, D. & Buchbinder, M. (2005) 'Community-based participatory research in South Africa: engaging multiple constituents to shape the research question', *Social Science & Medicine*, 61 (12), pp. 2577-2587.

Philipson, T.J. & Jena, A.B. (2005) *Who benefits from new medical technologies? Estimates of consumer and producer surpluses for HIV/AIDS drugs*. National Bureau for Economic Research.

Popay, J. & Williams, G. (1996) 'Public health research and lay knowledge', *Social Science & Medicine*, 42 (5), pp. 759-768.

Rosenfield, P.L. (1992) 'The potential of transdisciplinary research for sustaining and extending linkages between the health and social sciences', *Social Science & Medicine*, 35 (11), pp. 1343-1357.

Sachs, L. & Kranz, I. (1997) 'Interdisciplinary health research – a symposium', *Social Science & Medicine*, 44 (8), pp. 1209-1210.

Sewankambo, N.K. & Katamba, A. (2009) 'Health systems in Africa: learning from South Africa', *The Lancet*, 374 (9694), pp. 957-959.

U.S. National Library of Medicine (2017) Health statistics. MedlinePlus. Bethesda.

Waraich, P., Goldner, E.M., Somers, J.M. & Hsu, L. (2004) 'Prevalence and incidence of mood disorders: a systematic review of the literature', *The Canadian Journal of Psychiatry*, 49, pp. 124-138.

World Health Organization (2018) Health statistics. Geneva.

Welman, J.C., Kruger, S.J. & Mitchell, B.C. (2005) *Research methodology*. 3rd ed. Cape Town: Oxford University Press Southern Africa.

Chapter 8

Routine Data Sources and Descriptive Epidemiology

The topics of discussion in this chapter are the following:

- Using death certificate data in research
- Types of data, data summaries and correlation
- Demographic, health effect, population-based health data, and the ecological fallacy
- Scenario for calculating standardised mortality ratio

8.1 Using death certificate data in research

DEATH CERTIFICATE DATA can be of value for the purposes of epidemiological research, as epidemiological research helps us to practically make associations between information or data, and the risk factors for disease outcomes in groups or populations. Joubert & Ehrlich (2005) acknowledge that the notion of cause or risk factor could be tricky because most conditions require a number of causes or risk factors working together before the disease outcome could occur. It is important to realize that the validity and reliability of each data element has to be taken into consideration (Bruce et al, 2008).

According to Cohen et al (2007), death certificates contain important information such as age, sex, race, occupation, residential address, date and place of death which could be used in epidemiological studies for the purpose of investigating possible factors associated with breast cancer. The approach would be to identify a country and analyse the death certificate data for a specific period such as from year 2000 to year 2010, and then separate the information by age, sex, occupation, residential address, and place of death. One would conduct a case-control study, and my cases would be all deaths due to breast cancer and my controls would be all deaths other than cancer.

The strengths of death certificate data include its relative ready availability and relative low cost which would make it quicker to interpret the data; and its completeness would allow for interpretation and description of patterns of in-country and cross-country populations (Cohen et al, 2007). The other strength is that death certificate data can be used as a baseline, particularly in states or countries where the use of mammography is not yet well established (Geffken, 2000). Since all types of death are reported on death certificates, it implies that the interpretation of the data would have a lesser probability of being biased.

The limitations of death certificate data include the inconsistencies with which the variables are recorded in different states or countries (Geffken, 2000). Such variations would affect the interpretation of the data in the sense that one would not be able to confidently apply possible factors associated with breast cancer from one country to another country. Another limitation or weakness is the fact that the recording of the cause of death is largely dependent on the physician, because sometimes the physician would just record it as a medical condition instead of being specific. The recording of the cause of death as being a medical condition has somewhat become a norm, particularly in some of the sub-Saharan countries. This implies that using death certificate data to investigate possible factors associated with breast cancer could be a challenge. Hoel et al (1993) highlight the challenge posed by cancer mortality studies because of the issues of quality of the data.

Death certificate data presents a good opportunity to investigate possible factors associated with breast cancer. However, the strengths and limitations need serious consideration as these could have health policy implications.

There are implications of the accuracy or inaccuracy of the death certificate data on research and beyond research. Bruce et al (2008) acknowledge the international variations in the way data is recorded on death certificates. As an extension of this thread, one is curious to find out what is actually recorded on the death certificate in your respective countries or where you are currently based, if the person who has died has been infected with HIV and likely suffered from AIDS?

The foregoing question is being asked because there is on-going debate in parts of sub-Saharan Africa regarding this very point. One group suggests that it should be recorded that the person has died of AIDS, while the second group suggests that the specific disease such as tuberculosis, breast cancer, and pneumonia, among others, should be recoded. As if this was not enough, there is another point of view coming from an ethical angle, which suggests that neither AIDS nor the specific disease should be recorded, but prefers that medical condition or natural causes be recorded instead. The recording of medical condition or natural causes seems to be widespread in the under-resourced public health sector of some of the southern African countries, and this sector caters for the majority of the population (van Rensburg et al, 2008).

A couple of years ago, there was a leading story that flooded the South African media, and this story was about the claim that some doctors in one of the South African provinces were actually recording that patients were suffering from AIDS and therefore died of AIDS. Interestingly, this story raised a huge public outcry in a country that is reeling from the brunt of the epidemic, where it has been reported that

HIV and AIDS are a common denominator for the high mortality, and particularly for the very high maternal mortality and child mortality in South Africa (Pillay, 2008). It is very striking that while South Africa accounts for approximately 0.7 per cent of the world's population, it has more than 17 per cent of the world's new HIV infections (Vecchiatto, 2011).

The other dilemma is that there are people who die at home without having attended any clinic or hospital, and this also raises many questions about what to accurately record on the death certificate. It is interesting to note that it has now become acceptable practice in a number of developed countries to allow terminally ill patients to die at home, and hopefully this does not create issues with recording the correct reason for death (Cohen et al, 2010).

The dilemma of what should be recorded on death certificates poses huge challenges for epidemiological research, and this stifles possible opportunities for intervention, particularly in Southern Africa where the ravages of HIV and AIDS are so brutal.

The mobility of people has increased in recent times. The increase in the mobility of people is possibly fuelled by the fact that the world has now become a global village (Hill, 2009). We see a lot of multinational companies expanding into many corners of the world in a bid to gain a competitive advantage and sometimes in the name of foreign direct investment. As these expansions happen, there is importing and exporting of skills which is accompanied by frequent travels of people from one place to another. Tourism and education are some of the other major reasons why people frequently move from one place to another. What has this got to do with the recording of data on death certificates?

Well, this has much to do with the recording of data on death certificates in the sense that there is a possibility that the diagnosis can be missed if a person who has contracted a disease that is unique to the region where or she comes from, travels to another region where such a disease is not normally known or dealt with. The reality is that that new destination may not even have the tools or tests to confirm that disease. In such an instance, there is a high possibility that a person may be misdiagnosed, and when that person dies there, the wrong diagnosis would very likely be recorded on the death certificate. Cohen et al (2007) emphasise the need for accuracy of information on death certificates to ensure that epidemiological research produces meaningful results that could benefit health systems.

It once happened a few years ago whereby an international tourist who had just visited one of the countries in central Africa fell ill when he arrived in South Africa and was admitted in one of the hospitals in the Gauteng province of South Africa. The person subsequently died within a few days, and it was thought that he had died of other weird causes. The diagnosis of Ebola fever was completely missed despite the battery of tests that were conducted. It was only after a number of the hospital's staff members also fell ill and died showing similar symptoms to the person who had just died that the diagnosis of Ebola fever was finally made. The reason for the misdiagnosis was probably due to the fact that Ebola fever is very rare in South Africa (Nzimande, 1996). The other issue is that misdiagnosis may also be a case of underreporting. Bruce et al (2008) observe that doctors are surprisingly not keen to complete notification forms in most instances of notifiable infectious diseases.

In a nut shell, the mobility of people could contribute to misdiagnoses and recording of the wrong cause of death on death certificates. Misdiagnosis and underreporting can affect the completeness of data recorded on death certificates (Berrino, 2003).

There is prominence of the many variations with regard to the recording of death certificate data. In the midst of these variations and inconsistencies, there is an opportunity and a challenge for epidemiological research. We can still learn something from epidemiological research and we can enhance the opportunity for epidemiological research by ensuring that we pay more attention to the specific research question and by employing the appropriate study design (Bruce et al, 2008).

Evans (2003) mentions some of the study designs commonly used in epidemiology, and these include: cohort, cross-sectional, case control, and randomized control trials. Bruce et al (2008), highlight that we need to be aware of the strengths and weaknesses of each study design to ensure that it is the best for that particular study. Geffken (2000) prefers employing a case-control study to investigate the risk factors associated with breast cancer. It is believed that sometimes, a particular research question could be answered by using more than one study design (Joubert & Ehrlich, 2010).

Having said that, it has to be acknowledged that the best study design, on its own cannot solve the plethora of problems such as misclassification and underreporting which could result in wrong data being recorded on death certificates, as has already been highlighted earlier. In fact, even if the best epidemiological research method is employed, the outcome would not be of great value if the data on death certificates is inaccurate.

In retrospect, there is some positive that we can draw out of epidemiological research in the sense that it can create a stimulus for interventions aimed at addressing the risk factors associated with specific causes of morbidity or mortality in groups or populations. Such stimulus could act as a vehicle to heighten the authorities' focus on the health challenges at hand, thereby creating an opportunity for finding the root cause of the reasons for misdiagnosis or underreporting. If the death certificate data were accurate, it is possible that a pathway or pathways for breast cancer could be hypothesised. Sanderson et al (2006) observe that once a hypothesis has been formed about the pathways for breast cancer, the risk of dying from breast cancer could be influenced. Unfortunately, the disease pathways such as in cancer are not always obvious.

Epidemiological research has a role to play in terms of utilizing death certificate data or any other data source for the purpose of investigating possible factors that are associated with a particular disease. However, epidemiological research could play an even much greater role if the death certificate data were accurate. The reality is that there is great inconsistency and variation in the way death certificate data is recorded and utilised within a country and across countries. This is a much deeper challenge, particularly in many parts of sub-Saharan Africa where there are no clear standards about recording data on death certificates.

8.2 Types of data, data summaries and correlation

TYPES OF DATA

Variable	Type
Age	Continuous
Gender	Categorical
Ethnic group	Categorical
Self-rated health on 5-point scale ('very poor' to 'excellent')	Ranked/ordinal
Height	Continuous
Weight	Continuous
Satisfaction with health care (rated on 10-point scale)	Ranked/ordinal

Summarising data

- Calculating the mean and median for men:

Mean = Sum of scores/Number of scores (Christensen, 2007)

Given sum of scores = 19+30+70+77+46+41+74+53+69+68 = 547

Number of scores = 10

Therefore, the mean = 547/10 = 54.7

According to Joubert & Ehrlich (2007), the median is the 50th percentile or the value which divides the distribution in half whereby the values are arranged from small to large.

Given values arranged from small to large: 19 30 41 46 **53** **68** 69 70 74 77

Therefore, the median = (53 + 68)/2 = 60.5

This mean tells us that the average of the distribution is 54.7 and the median tells us that half the distribution falls above 60.5 and half falls below 60.5

We can also see that the median is greater than the mean, which implies that the distribution is left skewed because more than 50 per cent of the distribution is more than the mean (Bruce et al, 2008).

- Calculating the mean and median for women:

Mean = Total of individual scores/Number of scores (Welman et al, 2010)

Given sum of scores = 94+94+70+63+57+37+71+90+70+60 = 706

Number of scores = 10

Therefore, the mean = 706/10 = 70.6

Median is the midpoint in a distribution, which implies that half the scores fall above the median and half fall below it (Mouton, 2002).

Given values arranged in order: 37 57 60 63 **70 70** 71 90 94 94

Therefore, the median = (70 + 70)/2 = 70

This mean tells us that the average of the distribution is 70.6 and the median tells us that half the distribution falls above 70 and half falls below 70.

We can also see that the median is almost equal to the mean, because there is a difference of only 0.6 which implies that the distribution is almost uniform but very slightly skewed to the right because the mean is only 0.6 more than the median. However, if we do not consider this slight difference of 0.6, we could say the distribution is symmetrical because 50 per cent of the distribution would be more than the mean and the median, and 50 per cent would be less than the mean and the median (Bruce et al, 2008).

Correlation

- From the information in the scatterplot, we can say there is a strong positive relationship between child height and weight.
- Given the correlation coefficient: $r = .96$

The coefficient of determination $= r^2 = (.96)^2 = .92$

This implies that 92 per cent of the total variation in weight is explained by the variation in height. However, there is still a remainder of 8 per cent of the variation which may not be easily explained because it may be due to unknown factors (Bruce et al, 2008).

8.3 Demographic, health effect, population-based health data, and the ecological fallacy

THE ISSUE OF INFLUENZA vaccinations really needs serious attention, as some of the research studies have highlighted that the influenza vaccine does not confer protection to approximately 42 per cent of the studied population. Of course, we may not draw conclusions based on one study or a few studies, but the fact that some people got sick after the vaccinations were administered is an issue, nevertheless. In such cases, it makes it even more important for us to pay particular attention to the weaknesses of our research studies to ensure that we do not put people at risk at the cost of research. Although some researchers indicate that influenza vaccinations are very crucial for people that are deemed to be vulnerable, such as children and the elderly, other researchers report that the effectiveness of these vaccinations has not been convincing (Katz, 1987). Talbot et al (2010), add that currently, there is controversy about the immunogenicity of seasonal trivalent influenza vaccine in certain populations.

Given this situation, one wonders if the consistent worldwide recommendation to vaccinate people, particularly those that are deemed vulnerable such as children and the elderly is a fallacy (Rida et al, 2011). An observation is that the controversies take different forms in different contexts (Lantos et al, 2010). In the more affluent areas such as in the developed countries, parents tend to have fears about negative consequences of the vaccinations on their children, whereas in the rural areas where the majority of the people live, parents tend not to have many questions.

A couple of years ago, it once happened that there was a massive project in collaboration with an international health/development agency to administer an anti-helminth vaccine on all primary school children in one of the southern African countries. The first phase of the project yielded dire results as many of the vaccinated children fell ill and had to be admitted into hospitals and a few of them were subsequently reported to have died as a result of the anti-helminth vaccine. This was a big story in the local media, and parents were also raising their disapproval of such mass immunization as the immunization was done while the children were at school. The government of that particular country acted quickly, and ordered that the immunization be stopped with immediate effect.

What one is trying to highlight from that southern African country's story is that an ecological study was probably done, and this might have indicated that the country's children were at high risk of helminth infections. However, it was probably a fallacy because it assumed that all individual children of that country shared the same risk and characteristics. Sabatelli et al (2008), highlight that the dynamics of infections are very complex as they may be influenced by many factors arising from a variety of exposures and immunity-related factors. Therefore, the unique variations in different places should be taken into account (Bruce et al, 2008).

A colleague highlighted the issue of polio in Nigeria and the vulnerability of children who live in low-socioeconomic settings. It is possible that the broad-brush global immunisation of all children under the age of five against polio is a fallacy. It has now been discovered that after all these many years of polio immunisation, the OPV carries some risk in the sense that it is excreted and can circulate in the environment and thereby result in vaccine derived polio virus (vDPV) or a wild type of polio virus (Davies, 2013). Although there have only been approximately 114 virologically confirmed vaccine derived polio cases worldwide thus far, it is one too many (Davies, 2013).

At least the World Health Organisation has acknowledged the risk of vaccine derived polio virus, and has advised that a combination of the oral polio vaccine (OPV) and the inactivated polio vaccine (IPV) be used in the interim, with the aim of phasing out the OPV in the next couple of years. Since the IPV is more expensive than the OPV, the cost implications will be a challenge for many countries, particularly developing countries as this combination of OPV/IPV will cost much more. If it cost $600 000 for the project to administer an OPV to 650 000 children in Madagascar between August 2005 and October 2005, one can imagine the cost implications of the OPV/IPV combination (UNICEF, 2005). The draft Polio Eradication and Endgame Strategic Plan (2013-2018) suggests that at least one dose of IPV should be universally adopted as part of routine immunization by 2015 (Polio Eradication Initiative, 2013).

Although Nigeria is the only polio-endemic country in Africa, the eradication of polio still remains an imperative for all African countries

(Cooke, 2012). It is encouraging to see that this imperative has been shown by African governments' resolution through the African Regional Committee declaring that the persistence of polio is a national public health emergency (Kakujaha, 2011). This implies that all African countries should scale up their efforts in ensuring that polio is eradicated. I say this because there have been at least ten African countries that have reported cases of polio during the past few years, and it has been shown that these have generally been cases of imported polio (UNICEF, 2011).

Much as we have seen that there has been a declaration by the African Regional Committee for the eradication of polio, we need to bear in mind that high level political endorsement does not always translate into effective implementation on the ground. Well, perhaps, the high level political endorsement on immunization is also a fallacy. Nevertheless, it has to be acknowledged that immunisations are one of the greatest public health achievements of the 20th century (Vangjel, 2011).

It would be interesting to conduct an ecological study with particular focus on assessing readily available data, on the aggregated mean blood pressure of shift workers in the agricultural, manufacturing and mining sectors of southern African countries such as the Kingdom of Swaziland (Eswatini), Republic of South Africa, Botswana and Namibia, among others. The 'healthy worker effect' will certainly be taken into consideration as this data will only be focused on employees and not the general public. In fact, the 'healthy worker effect' could be one of the reasons the blood pressure means may be underreported in some instances, as highlighted earlier.

If this ecological study indicates a strong association between shift-work and high blood pressure, it may be tempting to conclude that it is a causal association (Bruce et al, 2008). It may also be tempting to stretch this association to other cardiovascular disorders. For that reason, the title of the study below should have been more specific to high blood pressure rather than cardiovascular disorders as a whole because there is a plethora of cardiovascular disorders that may not necessarily be linked to high blood pressure. Be as it may, the ecological study could enable us to generate a hypothesis about the association or no association between shift-work and high blood pressure. This could then be investigated further through more specific and in-depth study designs to see if there is a causal effect or no causal effect at an individual level. If the more specific studies find no causal association between shift-work and high blood pressure at individual level and yet the ecological study showed a strong association, it implies that this could be an ecological fallacy.

Further studies could also help us to identify possible other factors that could influence the causal association between shift-work and high blood pressure at individual level, if there is any at all. In other words this tells us that an ecological study can pave the way for more in-depth investigations about causal relationships between variables. However, this does not by any means imply that the in-depth studies do not have their own limitations. Since this study will be an international study, some researchers highlight that confounding is more pronounced in international studies, and this could be a major weakness as countries sometimes differ markedly in many aspects (Althabe et al, 2006).

Moreover, given the migration of workforce within the Southern Africa region, it is possible that the sectors of focus in this study will have varying degrees of heterogeneity which could influence the data. Pickett & Pearl (2001) highlight that the bigger the degree of heterogeneity, the bigger the margin of error.

CARDIOVASCULAR DISORDERS among shift workers: an ecological study of Southern African countries

Introduction:

The health effects of shift work have been an issue of debate among researchers. There is paucity of studies on the health effects of shift work among industrial workers in Southern Africa. The majority of shift workers in the agricultural, manufacturing and mining sectors of Southern Africa are blue collar and they come from low socio-economic backgrounds (Naidoo, 2009). Sub-Saharan Africa is now faced with an epidemic of non-communicable diseases such as high blood pressure (Mayosi et al, 2009). There is a possibility that shift work causes cardiovascular disorders (LaDou, 2007). The objective of this ecological study is to test the hypothesis that shift work causes high blood pressure among shift workers.

Methodology:

The cross-sectional ecological study design, will involve an assessment of annual occupational health reports of the agricultural, manufacturing and mining sectors of the Kingdom of Swaziland (Eswatini), Republic of South Africa, Botswana, and Namibia, to determine the mean blood pressure of shift workers in the respective industries. The annual occupational health reports to be assessed will be for the period 2010 to 2013.

Discussion:

The ecological fallacy might influence one's interpretation of the study results in the sense that these results will not be a true reflection of each individual represented by the mean blood pressure data in this study (Pickett et al, 2005). There is a possibility that there are other confounding factors that might be uniquely collateral or contralateral for the health outcome of blood pressure for each individual despite the similar exposure which in this case is shift work (Macinko, et al, 2003). Genetic factors may come into play as well as individuals have different predispositions to cardiovascular disorders such as high blood pressure than others. It is also important to consider that some of the employees in the industries of focus in this study may have migrated from other countries, and as a result, their predisposition may also be unique. Schwartz (1994) observes that there is complexity of the interplay of aetiological factors of any particular disease.

Moreover, the shift pattern may differ from industry to industry and from country to country, which implies that it would not be proper to conclude with certainty about whether shift work has direct effect on an individual's blood pressure or not. Bruce et al (2008) observe that sources of data do not provide the full picture.

The 'healthy worker effect' cannot be ruled out, because it is possible that there could be underreporting of the mean blood pressure as sickly workers with uncontrolled high blood pressure or with complications from high blood pressure may have been taken out of service or boarded off. Pearce (2000) argues that much as we have to be aware of the 'ecological fallacy' we also need to be careful of the 'individualistic fallacy'.

Conclusion:

An ecological study can be useful as a quick 'scan' to give us an idea about a possible association between an exposure and a health outcome in populations (Pearce, 2000). However, such a study has its limitations as it is prone to the ecological fallacy which assumes that all the employees of the industries that were assessed in this study share the social and health characteristics of that area.

8.4 Scenario for calculating standardised mortality ratio

STANDARDISATION

- Prior to 31 March 2013, and going as far back as 2001, there existed a number of Primary Care Trusts (PCTs) as part of the National Health Service (NHS) in England. Up to 2011, these PCTs provided community services directly, and their total budget was approximately 80 per cent of the total NHS budget. As such, the PCTs served a sizable population. However, it was not appropriate to compare the crude mortality rates within the PCT population with those for England and Wales because this could have been influenced by the difference in age distribution between the PCT population and that of England and Wales. For example, it may happen that the PCT population in Bolton had a much higher proportion of older people compared to other PCTs, and this could be masked if we compared with the age proportions of England and Wales, in general. Joubert & Ehrlich (2010) report that older age groups tend to have a higher mortality rate compared to younger age groups.
- Calculation of a standardized mortality ratio (SMR) makes it appropriate to compare mortality rates in this situation because the SMR takes account of the differences in age distribution of the PCT population and that of Wales and England. Heijink et al (2008) report that calculating a standardized mortality ratio compensates for the differences in the distribution mix.
- The SMR is an indirect form of standardization
- Calculation of the expected deaths for the Bolton PCT:

According to Joubert & Ehrlich (2010), the expected deaths for a specific age group in specific population can be calculated by multiplying the 'age-specific death rate of standard population by the population size of matching age category of the population of interest. Adding these can then give the 'expected deaths' of the population of interest (Bruce et al, 2008)

Based on the above explanation and the given table below, we can calculate the 'expected deaths' in Bolton PCT as follows:

'Expected deaths' for age group (0-4) = 1.61/1000 x 950 = 1.53

'Expected deaths for age group (15-24) = 0.98/1000 x 724 = 0.71

'Expected deaths for age group (25-44) = 2.32/1000 x 932 = 2.16

'Expected deaths for age group (45-64) = 13.64/1000 x 870 = 11.87

'Expected deaths for age group (65-84) = 64.45/1000 x 423 = 27.26

'Expected deaths for age group (85+) = 187.13/1000 x 24 = 4.49

Therefore,

Expected deaths for the Bolton PCT = 1.53 + 0.71 + 2.16 + 11.87 + 27.26 + 4.49 = **48.02**

Age group	Deaths in Bolton PCT	Bolton PCT population	England and Wales mortality rate (per 1,000/year)	Deaths 'expected' in Bolton PCT
0-14	2	950	1.61	1.53
15-24	1	724	0.98	0.71
25-44	5	932	2.32	2.16
45-64	12	870	13.64	11.87
65-84	26	423	64.45	27.26
85+	8	24	187.13	4.49
Total	54	3,923	12.41	**48.02**

- According to Bruce et al (2008),

SMR = Total number of observed deaths / Total number of expected deaths

Therefore, SMR for the Bolton PCT = 54 / 48.02 = 1.12

It is noteworthy that the reference SMR is usually expressed as 100 (Brown & Barraclough, 2000). This implies that the SMR for the Bolton PCT can be expressed as 1.12 x 100 = **112**

Based on the SMR for the Bolton PCT, it can be interpreted that independent of the influence of age distribution, the SMR for the Bolton PCT is 12 per cent higher than that of England and Wales (Bruce et al, 2008).

- Calculating and interpreting the confidence interval for the SMR:

Upper limit of 95% CI for SMR = 112 + [1.96 x (112/ square root of 54)]

= 141.873

Lower limit of 95% CI for SMR = 112 – [1.96 X (112 / square root of 54)]

= 82.127

We can say that the Bolton PCT SMR is 112, with a 95 per cent confidence interval of 82 – 142. This implies that we can be 95 per cent certain that the true SMR for the Bolton PCT lies between 82 and 142, or 18 per cent below and 41 per cent above the England and Wales' level of mortality (Bruce et al, 2008). The confidence interval of 82 – 142 is wide. Maree (2010) highlights that there is a higher level of confidence if the confidence interval is wider, and there is a lower level of confidence if the confidence interval is narrower. In this case, we can say we have a higher level of confidence that the true SMR for the Bolton PCT lies between 82 and 142.

REFERENCES

Berrino, F. (2003) 'The EUROCARE study: strengths, limitations and perspectives of population-based, comparative survival studies', *Annals of Oncology*, 14 (5), pp. v9-v13.

Brown, S. & Barraclough, B. (2000) 'Causes of the excess mortality of schizophrenia', *The British Journal of Psychiatry*, 177, pp. 212-217.

Bruce, N., Pope, D. & Stanistreet, D. (2008) 'Routine data sources and descriptive epidemiology'. In: *Quantitative research methods for health research: a practical guide to epidemiology*. Chichester: John Wiley & Sons, Ltd.

Christensen, L.B. (2007) 'Data analysis'. In: *Experimental methodology*. 10th ed. Boston: Pearson.

Cohen, J., Houttekier, D., Onwuteaka-Philipsen, B., Miccinesi, G., Addington-Hall, J., Kaasa, S., Bilsen, J. & Deliens, L. (2010) 'Which patients with cancer die at home? A study of six European countries using death certificate data', *Journal of Clinical Oncology*, 28 (13), pp. 2267-2273.

Cohen, J., Bilsen, J., Missinesi, G., Lofmark, R., Addington-Hall, J., Kaasa, S., Norup, M.,Van der Wal, G. & Deliens, L. (2007) 'Using death certificate data to study place of death in 9 European countries: opportunities and weaknesses', *BioMed Central Public Health*, 7, pp. 283-292.

Cooke, J.G. (2012) *Polio in Nigeria*. Centre for Strategic and International Studies [Online]. Available from: http://www.csis.org

Evans, D. (2003) 'Hierarchy of evidence: a framework for ranking evidence evaluating healthcare interventions', *Journal of Clinical Nursing*, 12, pp. 77-84.

Davies, R. (2013) 'Inactivated polio vaccine: its proposed role in the final stages of polio eradication', *The Pan African Medical Journal*, 14, p. 112.

Geffken, D.F. , Perry, M.J. & Callas, P.W. (2000) 'Association of occupation and breast cancer mortality in the state of Vermont, 1989-1993', *Massachusetts Journal of Medicine*, 5, pp. 75-79.

Heijink, R., Koolman, X., Pieter, D., Van der Veen, A., Jarman, B. & Westert, G. (2008) 'Measuring and explaining mortality in Dutch hospitals: the hospital standardization mortality rate between 2003 and 2005', *BMC Health Services Research*, 8, p. 73.

Hill, C.W.L. (2009) *International business: competing in the global marketplace.* 7th ed. Boston: McGraw-Hill International Edition.

Hoel, D.G., Ron, E., Carter, R. & Mabuchi, K. (1993) 'Influence of death certificate errors on cancer mortality trends', *Journal of the National Cancer Institute*, 85 (13), pp. 1063-1068

Joubert, G. & Ehrlich, R. (2007) *Epidemiology: a research manual for South Africa.* 2nd ed. Cape Town: Oxford University Press Southern Africa.

Kakujaha, A. (2011) 'Polio: much done, much still to be done', *The Southern Times*, 11 November.

Katz, S.L. (1987) Controversies in immunisation', *Pediatric Infectious Disease Journal*, 6 (6), pp. 607-613.

LaDou, J. (2007) Stress at work. In: *Occupational & environmental medicine.* New York: McGraw-Hill.

Lantos, J.D. Jackson, M.A., Opel, D.J., Maree, M. & Myers, A., Connelly, B.L. (2010) 'Controversies in vaccine mandates', *Current Problems in Pediatric and Adolescent Health Care*, 40 (3), pp. 38-58.

Macinko, J., Starfield, B. & Shi, L. (2003) 'The contribution of primary care systems to health outcomes within organisation for economic cooperation and development (OECD) countries, 1970-1998', *Health Services Research*, 35 (3), pp. 831-865.

Maree, K. (2010) 'Statistical analysis II: inferential statistics'. In: *First steps in research*. Pretoria: Van Schaik.

Mayosi, B.M., Flisher, A.J., Lalloo, U.G., Sitas, F., Tollman, S.M. & Bradshaw, D. (2009) 'The burden of non-communicable diseases in South Africa', *The Lancet*, 374(9693), pp. 934-947.

Mouton, J. (2002) 'Data analysis and interpretation'. In: *Understanding social research*. Pretoria: Van Schaik.

Naidoo, R. (2009) 'The work and health in Southern Africa (WAHSA) programme – overall experience and the way forward', *Occupational Health Southern Africa*, 15, pp. 2-6.

Nzimande, P.N. (1996) *Communicable in the African continent*. 2nd ed. Pinetown: Alberts Publishers.

Pearce, N. (2000) 'The ecological fallacy strikes back', *Journal of Epidemiology & Community Health*, 54, pp. 326-327.

Pickett, K.E., Kelly, S., Brunner, E., Lobstein, T. & Wilkinson, R.G. (2005) 'Wider income gaps, wider waistbands? An ecological study of obesity and income inequality', *Journal of Epidemiology & Community Health*, 59 (8), pp. 670-674.

Pillay, R. (208) 'Managerial competencies of hospital managers in South Africa: a survey of managers in the public and private sectors', *Human Resources for Health*, 6 (4), pp. 1-7.

Polio Eradication Initiative (2013) *Polio eradication and endgame strategic plan (2013-2018)*.

Sabatelli, L, Ghani, A.C., Rodrigues, L.C., Hotez, P.J. & Brooker, S. (2008) 'Modelling heterogeneity and the impact of chemotherapy and vaccination against human hookworm', *Journal of the Royal Society Interface*, 5 (28): pp. 1329-134.

Sanderson, M., Daling, J.R., Doody, D.R. & Malone, K.E. (2006) 'Perinatal factors and mortality from breast cancer', *Cancer Epidemiology, Biomarkers & Prevention*, 15, pp. 1984-1987.

Schwartz, S. (1994) 'The fallacy of the ecological fallacy', *American Journal of Public Health*, 84 (5), pp. 819-824.

Talbot, H.K., Rock, M.T., Johnson, C., Tussey, L., Karita, U., Shanker, A., Shaw, A.R. & Taylor, D.N. (2010) 'Trivalent influenza vaccine when given with vax102, a recombinant influenza m2e fused to the tlr5 ligand flagellin', *PLoS ONE*, 5 (12), pp. 1-7.

UNICEF (2011) *Polio eradication in Eastern and Southern Africa*. Available from: http://www.unicef.org/esaro/5479_polio.html

Vangjel, L. (2011) '*The economics of child health*', Humanities and Social Sciences, 72 (2-A): p. 686.

Van Rensburg, D.H.C.J., Steyn, F., Schneider, H. & Loffstadt, L. (2008) 'Human resource development and antiretroviral treatment in Free State province, South Africa', *Human Resources for Health*, 6 (15), pp. 1-10.

Vecchiatto, P. (2011) *Current health system not sustainable*. Available from: http://www.businesslive.co.za/southafrica/2011/06/29/current-health-system-unsustainable

Welman, C., Kruger, F. & Mitchell, B. (2010) 'Data analysis and interpretation of results'. In: *Research methodology*. Cape Town: Oxford University Press.

Chapter 9

Surveys

This chapter covers the following topics:

- Survey design
- Methods of sampling

9.1 Survey design

IMAGINE A RESEARCHERS' population of interest being all the people who live in Aigburth (a suburb of Liverpool). The first step would be to have a sampling frame which includes the characteristics of the sample such as the name and surnames, contact details, addresses, telephone numbers. The sampling frame could be obtained from a number of sources such as telephone directory, municipality register, or even lists from general practitioners in that suburb. Although it is the ideal to get the list of the entire population of Aigburth, but we have to realize that the sampling frame may not contain the list of every person in that suburb.

The method of sampling that could be used is simple random sampling, whereby each person on the list will be given a specific code such as A1, A2, and so on. Suppose the researchers want to choose 1000 out of a total of 5 000 people in the sample frame, then 5000 codes will be randomly selected for example by using a computer programme. It has to be borne in mind that simple random sampling is dependent on chance. This means that, if we are lucky, we may get a sample that is a true representation of the population of interest, but if we are not lucky, the sample may not be a true representation of the population of interest. It is worth noting that a sampling error may occur if different random samples of the same size such as 1000 from the same sample frame give different means which may also differ from the mean of the population of interest.

Once the researchers have confirmed the randomly selected sample from the sampling frame, they should conduct a survey through the use of questionnaires so that they could estimate the proportion of people in Aigburth who regularly smoke. Before the survey is conducted, it would be important to make the participants aware about the research survey, in advance, and this should be clearly explained to them so that they are prepared. The questionnaires with a cover letter can then be mailed to the individuals' addresses. The cover letter should clearly explain what the survey is all about and the deadline for submitting the completed questionnaires should be clearly indicated. It is possible that not all questionnaires will be completed or returned due to reasons such as inability, relocation, among others.

The implications of the sample size are that there could be a possible underrepresentation of the population and an overrepresentation of the population. It is therefore important to calculate the sample size with some degree of precision to ensure that there is accuracy in the sample mean which should reflect a true picture of the population of interest. In this regard, the researchers may need to calculate the standard error and the 95 per cent confidence interval to estimate the proportion of people who regularly smoke to within a specified level of error.

It is generally believed that a larger sample size provides a better chance to yield information or results that could be generalised to the population of interest. It is indeed good to have a large sample size especially when we study larger populations (Maree, 2010). However, it may happen that the sample size could be unnecessarily large, and this could create many problems in terms of the collection and analysis of the large amounts of data and unnecessary costs could also be incurred. Joubert & Ehrlich (2007) observe that sometimes when we study larger populations it is a matter of knowing how large the sample size should be. Bearing that it in mind, it would probably be a better idea to involve a statistician to assist with calculating an ideal sample size and make the necessary adjustments according to the sampling method that we have chosen (Bruce et al, 2008). Such adjustments could indicate the design effect of the sampling method and this indicates how the sampling design influences the accuracy of the estimates (Joubert & Ehrlich (2007).

Some studies highlight that the sample size does not necessarily have to be large when a stratified sampling method is employed, and yet stratified sampling has a better chance of having a more representative sample than simple random sampling (Welman et al, 2005). To illustrate this point, suppose there is a total population of 5000 at Aigburth (a suburb of Liverpool) and the researchers wish to investigate the proportion of people who regularly smoke in that suburb. If the researchers decide to use simple random sampling, they may need to study a larger proportion of the population such as 1000 people. It may happen that the selected sample of 1000 randomly selected people predominantly comprises one segment of the population, and certain segments of the population may not be represented at all.

However, if the researchers prefer to use stratified sampling, they will have to ensure that at least all strata of the population are included (Welman et al, 2005). In other words, the researchers will ensure that they randomly select a similar percentage of people from each stratum of the population. In this regard, the different strata of the population of Aigburth could be the different age categories or the different sexes or the different ethnic groups.

As part of our self-care programme, we will possibly consider employing a stratified sampling method to estimate the proportion of people who regularly drink alcohol in different countries of Southern Africa. Cost implications and the amount of time required to conduct such a study will influence one's choice of the study design as well. Johnson et al (2001) observe that the willingness to participate and honesty with regard to how the respondents respond in a survey is another important factor that can influence the accuracy of the information to be analysed.

In survey research, the intention is to have accurate information that is representative of the population of interest (Bruce et al, 2008). However, getting accurate information that is representative of the population of interest has become one of the biggest challenges facing researchers. It is a truism that "one of the interesting things about an intended random sample is that those who actually return the questionnaires may not end up being a random sample" (Holmes, 2013). This indicates that researchers have to come up with ways of dealing with this problem.

Interestingly, some researchers have proposed various creative ways of dealing with this problem, but some of these various ways have not been effective in all settings. Grava-Gubins & Scott (2008) observe that conducting surveys among professionals remains problematic because the response rate does not improve even after implementing various creative ways such as shortening the survey, providing incentives, making follow-ups and creative marketing.

Some researchers report that employing a tailored design method yields positive results because it improves response rate in postal surveys (Thorpe et al, 2008). It is highlighted that the tailored design method involves key elements such as: personalized correspondence; using a respondent-friendly questionnaire; using return envelopes with first class registered mail; financial incentives; and special follow up contact by certified mail or telephone call (Scott et al, 2011). This could possibly work for populations that are in the developed countries and possibly also in the urban areas of developing countries, because the majority of people who live in developing countries such as in Southern Africa, are in the rural areas where mail correspondence could be a challenge. VanGeest et al (2007) report that the response rate among professionals increases if the survey research is endorsed by a professional body or association. Some studies indicate that surveys that have a focused participatory approach with incentives for participating show better response rates especially among the youth (Claudio & Stingone, 2008).

This highlights the importance of ensuring that a survey research should be tailor made for that particular population of interest as settings always have many variations in terms of demographics, culture and norms. In that regard, I do not think there is a standard way of dealing with the problem. For example, in many parts of Southern Africa, it would be crucial to get buy-in from the chiefs or councillors of that particular area of focus, before the survey research is even started. The chiefs and councillors are very influential in their respective communities, and once they have endorsed the idea of a proposed survey research, there is a better chance for a good response rate. A good response rate gives a better chance for a representative sample and accuracy of information, and minimizes the chance of a sampling bias.

Imagine a study aimed at measuring the prevalence of insomnia among the working population of the industrial city of uMhlathuze in the north east coast of the South African province of KwaZulu-Natal. The sampling frame would be obtained from the employee registers of all the different industries of the city of uMhlathuze. Stratified probability sampling would be employed, and the stratification would be based on age, sex, occupation, shift pattern, industry, and length of service. Joubert & Ehrlich (2007) observe that stratified probability sampling is easy to apply where full lists such as employee registers can be obtained.

An advance letter would be issued to all potential respondents explaining the purpose of the survey, explaining that participants would be selected at random, that participation would be voluntary, and that confidentiality would be maintained and guaranteed (Roth et al, 2011). The advance letter would also explain that participation in the survey would not affect the participants' employment, that participants would be free to opt out of the survey at any time or they could phone a specified toll-free number for further information, and that there would be a financial incentive for participating in the survey (Scott et al, 2011).

Consent would first be obtained from each participant before telephonic interviews are conducted (Roth et al, 2011). The telephonic interviews would be based on a brief insomnia questionnaire which would be a condensed hybrid of the World Health Organisation's International Statistical Classification of Diseases, Research Diagnostic Criteria for insomnia, International Classification of Sleep Disorders, and the DSM-IV-TR criteria (Roth et al, 2011). It is critical to pay attention to the question type, length of the questions, ordering of the questions, phrasing and structure of the questions (Bruce et al, 2008).

It is possible that some of the people randomly selected to participate in the survey may not contribute to the research, and this may negatively affect the response rate and the representativity of the population of interest. According to Bruce et al (2008), such people may include those that cannot be contacted and those that decline to participate in the survey. Sturgis et al (2006), highlight that the response rate in a survey is usually strongly linked to the characteristics of the environment of focus. Mosavel et al (2005) observe that the

South African population is very sensitive, and research surveys are usually met with suspicion from different angles, and in particular from a politico-socio-economic angle. The sensitivity of communities to research surveys is not unique to South Africa, but is a global issue, and the only difference is the magnitude of that sensitiveness.

We may need to come up with ways of gaining the trust of the population of focus in the industrial city of uMhlathuze, in order not to risk the possibility of a high non-response bias. It is recommended that stratum-specific results be proportionately weighted before the overall results can be analysed (Joubert & Ehrlich, 2007). It is worth mentioning that telephonic surveys can be tricky as any slight tension during the interview may lead to outright refusals (O'Rourke & Blair, 1983). In that vein, it is recommended that there be expertise in the planning, execution and analysis stages of such survey research to prevent or minimize non response bias (Draugalis & Plaza, 2009).

9.2 Methods of sampling

STRATIFIED RANDOM SAMPLING is indeed a good choice to make because it has two key advantages, including the fact there is a higher probability of a representative sample and that it requires a smaller sample size compared to simple random sampling (Welman et al, 2005). However, it is very important to ensure that stratum-specific results are proportionately adjusted or weighted when combining them especially if strata are not chosen proportionately to their size in the study population (Joubert & Ehrlich, 2007). It is considered that a constant sample fraction may be used to increase the level of precision in the sample estimate when a stratified sampling method is employed (Bruce et al, 2008).

The importance of ensuring that the researchers have a clear understanding of the composition of the population of interest, before the survey is even carried out, is of importance. It may happen that the sampling frame could be wrongly finalized, with the understanding that the demographic records of the entire geographical area of interest are complete, only to find that there is or are groups of people or strata that have been unwittingly excluded. There are a number of reasons why certain groups of people could be unwittingly excluded from the sample frame. These reasons may include: homelessness, illiteracy, living in informal settlements, and other various forms of social exclusion or marginalization. Igboamalu & Govender (2013) observe that illiteracy and low socioeconomic status have a huge bearing on low response rates in research surveys.

A typical example is that of South Africa where there are so many people living in informal settlements in the peri-urban areas and most of these people frequently migrate from one place to another in search for jobs. Another example is that of the indigenous minority groups such as the San of Southern Africa some of whom still lead a semi-nomadic life, and their language of communication is marginalized and not well understood in the mainstream (Ohenjo et al, 2006). In this regard, if we were to conduct a stratified random sampling based on demographic information obtained from municipality registers or telephone directories, a lot of the people who do not appear on the records could be missed. As a result this could lead to sampling error, even though sampling error is thought to be reduced in stratified random sampling compared to simple random sampling (Bruce et al, 2008).

It is suggested that a specific sampling method such as snowball sampling be employed in the case of challenges such as when dealing with people that are very difficult to reach (Bruce et al, 2008). It can be concluded that the different sampling methods have their advantages and disadvantages. It is therefore advisable that the research team choose a sampling method that is suitable for a specific situation or circumstance to ensure accuracy of data and analysis, and representativeness of the population of interest (Maree, 2010).

In survey work we use samples because it may not be possible to study the whole population (Joubert & Ehrlich, 2007). The sample should be a true representation of the population in which we want to base our research on (Grava-Gubins & Scott, 2008). For example, imagine the suburb of Aigburth has a population of 50 000 and we want to estimate the number of people who regularly smoke in that particular suburb. Using samples may not require that we study each and every individual in that suburb, but a portion of the population can be selected, and as long as the selected sample is representative of the general population of Aigburth, the number of people who regularly smoke can be estimated.

Selection bias means that we systematically choose a sample that is an underrepresentation or an overrepresentation of the population that we want to study (Bruce et al, 2008). In that case, it means that we cannot infer the results we get from the selected sample to the general population of interest. For example, a selection bias may occur if the selected sample differs from the general population of interest (Woolf et al, 2000).

Random sampling overcomes selection bias because in random sampling the selection of the sample is purely based on chance (Bruce et al, 2008). In random sampling, there is generally no subjective preference of a specific group in the population of interest. Although it cannot be ruled out that there is a chance that the randomly chosen sample may not be representative of the general population, but such a probability is small compared to when a non-random sampling is employed. Robins et al (2000) observe that the selection bias on un-observables nor its magnitude is minimized if random sampling is employed in survey research.

A sampling frame is a list that contains the full details of the individuals that we want to study in that population (Bruce et al, 2013). Whittmore & Knafl (2005) defines a sampling frame as the list of units to be studied. An example of a sampling frame is all the residents of a particular town whose details can be obtained from a voters register. The details on the voters register may contain the name, surname, date of birth, physical address, postal address, and telephone number of each resident who qualifies for inclusion in the survey research (Thorpe et al, 2008).

Standard error and confidence interval for population mean:

Standard error = s / square root of n

Given, n = 300

Given, s = 10.5

Therefore, standard error = 300 / square root of 52.3 = 1.45

95 per cent confidence interval around the sample mean:

(Mean – 1.96s / square root of n, mean + 1.96s / square root of n)

(52.3 – 1.96 x 10.5 / square root of 300, 52.3 + 1.96 x 10.5 / square root of 300)

Therefore 95 per cent confidence interval around the mean is (51.1, 53.5)

This implies that we are 95 per cent confident that the mean body weight of secondary school children lies between 51.1 kg and 53.5 kg.

It is true that stratification is usually employed for a specific reason. It is reported that stratification can be specifically employed to deal with confounding in the study design, in the analysis, or in both (Bruce et al, 2013). Confounding occurs when the exposed and non-exposed groups in the population of interest are not comparable due to inherent differences in background disease risk, usually due to exposure to other risk factors or extraneous factors (Cromar et al, 2006).

Suppose the aim of our research was to see if there is an association between shift work and insomnia among the working population of the industrial town of uMhlathuze in the South African province of KwaZulu-Natal. It is possible that we may get results indicating that not all the workers who were exposed to the same shift work, suffer from insomnia. The health outcome may not be a direct result of shift work, but it could also be influenced by other factors such as age, sex, smoking, alcohol consumption, and other inherent predispositions to insomnia.

In our study design, if we suspect alcohol consumption to be a confounding factor, we could stratify on alcohol consumption by dividing the population of interest into alcohol consumers and non-alcohol consumers, and we could select a specified proportion of cases and controls from each stratum (Bruce et al, 2008). It is highlighted that it is also possible to carry out post-stratification or stratification during the analysis of results even if the sample was not stratified, provided information on the extraneous factors was collected (Bruce et al, 2013). However, the challenge with post-stratification is that it cannot be guaranteed that there would be enough cases and controls in the different strata (Bruce et al, 2013). Nonetheless, by using a stratified analysis, the researcher attempts to adjust for, or remove the effect of, an extraneous factor which is related to the exposure and the health outcome (Joubert & Ehrlich, 2007).

Stratification is usually employed for a specific reason of eliminating confounding such as in case control studies (Bruce et al, 2008). Confounding elements could be used in the broad sense to include researcher, participant and context effects, and these pose a threat to the reliability of data collected (Mouton, 2002). It is therefore important to control for extraneous factors or confounding factors in order to attain internal validity (Christensen, 2007).

It is acknowledged that using stratification with more than two variables, may result in the sample getting smaller and smaller. It can be observed that having a shrinking sample could increase the chance of sampling bias. In that case, simple random sampling could be preferred.

However, it is also possible that using simple random sampling could result in a non-representative sample as those randomly selected could end up not being a random sample due to the likelihood of self-selection (Draugalis & Plaza, 2009). At the same time, since the sampling frame of the working population of the industrial city of uMhlathuze is so huge, simple random sampling could require a larger sample which could be very difficult to manage especially in the circumstance of budget constraints (Welman et al, 2005). It is interesting to note that some researchers report that smaller samples may not necessarily represent sampling bias, but what matters the most in survey research is the quality of how the sample is selected, and how the data is collected and analysed and to ensure that it is representative of the population of interest (McCarty, 2003).

In that regard, it is therefore tempting to use stratification, because in stratification we do not necessarily have to deal with a larger sample and we can also use a constant sampling fraction to increase the precision of our sampling estimate unlike in simple random sampling (Joubert & Ehrlich, 2007). Of course, it does make a lot of sense to stick to one variable or two variables when a stratified probability sampling is employed as this could help us avoid the risk of a shrinking sample size as when using more variables. Sticking to one variable such as shift pattern will also make it easier for us to manage in terms of analysis of the results as opposed to when more variables were involved. Bruce et al (2008) add that using stratified probability sampling is an advantage over using simple random sampling because stratification gives a better chance of a representative sample and can yield more reliable results.

Simple random sampling is preferred when more than two variables are used, and stratified probability sampling could be employed when a maximum of one or two variables is used. Variables such as age, sex, occupation or social class could still assist us in terms of assessing non-response bias, to see if there is a match between the responders and non-responders in the population of interest (Bruce et al, 2008).

The issue of antenatal care and perinatal outcomes is pertinent in every country, and, particularly so in sub-Saharan Africa, as this region is faced with a huge burden of disease, high crude birth rates, and also very high maternal and infant mortality rates (de Haan, 2005). Studies highlight that the maternal and infant morbidity and mortality issues are so serious in sub-Saharan Africa that such issues threaten to overwhelm the already weak health systems and have undermined the potential of this region to meet most of the targets of the Millennium Development Goals (Chopra, 2009). As such, if these issues are not addressed with a sense of urgency through effective leadership, achievement of the Sustainable Development Goals will also be a challenge.

Be as it may, we cannot afford to lose any hope because through epidemiological research, we can get an idea of the magnitude of possible associations with the health outcomes so that interventions could be implemented (Bruce et al, 2008). One such glimmer of hope was the then proposed cohort survey to explore the associations between background and behavioural decisions on antenatal care attendance and perinatal outcomes in Kenya. It was interesting to note that in the cohort survey, the plan was to recruit some people from the same communities where the research would be conducted so that they could assist with communicating the intended research message in the language that the target population was familiar with. That was a good idea, and that could assist in getting a good response rate because by involving members of the communities in conducting the research, there is a good chance of gaining the buy-in and trust of the population of interest for your research.

However, there could also be a downside in selecting certain members of the communities to assist with the research because this could potentially be viewed as a hidden agenda by some, especially if the target population is deeply divided or not homogeneous. I see such a challenge in many parts of the middle- and low-income countries where politics and/or religion and/or ethnicity and/or social class seem to stifle progress in many aspects, especially if the progress is initiated from the "other side" of the political, religious, ethnic or social class divide. Sometimes, the divide is directly or indirectly perpetrated by some of the high-income or donor countries who have their own interests in the middle- and/or low-income countries.

Perhaps, the best way to win the trust of polarised communities in the middle- and low-income countries is to begin by getting endorsement of the traditional leaders who tend to play a neutral and unifying role in such communities. Getting buy-in of the traditional leaders would not only help in getting a good response rate in the survey, but it could also help with effective and efficient implementation, ownership and sustainability based on the survey findings and recommendations. A good example is that of the involvement of traditional leaders and local councils in Uganda which made a huge difference in the fight against HIV and AIDS, and this has made Uganda a good model in this regard (Allan & Heald, 2004).

Maximum participation of the target population when we conduct survey research augers well for a good response rate and sustainable implementation of recommendations. It is important to have an approach that is tailored for that particular target population as each and every environment is unique.

REFERENCES

Allan, T. & Heald, S. (2004) 'HIV/AIDS policy in Africa: what has worked in Uganda and what has failed in Botswana?' *Journal of International Development*, 16, pp. 1141-1154.

Bruce, N., Pope, D. & Stanistreet, D. (2008) 'Surveys'. In: *Quantitative research methods for health research: a practical guide to epidemiology*. Chichester: John Wiley & Sons, Ltd.

Claudio, L. & Stingone, J.A. (2008) 'Improving sampling and response rates in children's health research through participator methods', *Journal of School Health*, 78 (8), pp. 445-451.

Chopra, M., Lawn, J.E., Sanders, D., Barron, P., Abdool-Karim; Q., Flisher, A.J., Mayosi, B.M., Tollman, S.M., Churchyard, G. & Coovadia, H. (2009) 'Achieving the health millennium development goals for South Africa: challenges and priorities', *The Lancet*, 374 (9694), pp.1023-1031.

Christensen, L.B. (2007) 'Reliability and validity in experimental research'. In: *Experimental methodology*. Boston: Pearson

Cromar, N., Cameron, S. & Fallowfield, H. (2006) 'Environmental epidemiology'. In: Environmental health in Australia and New Zealand. Melbourne: Oxford University Press.

De Haan, M. (2005) 'Health needs through the lifespan'. In: *The health of Southern Africa*. Cape Town: Juta Academic.

Draugalis, J.R. & Plaza, C.M. (2009) 'Best practices for survey research reports revisited: implications of target population, probability sampling, and response rate', *American Journal of Pharmaceutical Education*, 73 (8), pp. 142-149.

Grava-Gubins, I. & Scott, S. (2008) 'Effects of various methodologic strategies: survey response rates among Canadian physicians and physicians-in-training', *Canadian Family Physician*, 54 (10), pp. 1424-1430.

Igboamalu, C. & Govender, I. (2013) 'Contraceptive knowledge and practices among pregnant females in Lower Umfolozi District War Memorial Hospital, KwaZulu-Natal', *Occupational Health Southern Africa*, 19 (1), pp. 26-31.

Johnson, A.M., Mercer, C.H., Erens, B., Copas, A.J., McManus, S., Wellings, K., Fenton, K.A., Korovessis, C., Macdowall, W., Nanchabal, K., Purdon, S. & Field, J. (2001) 'Sexual behavior in Britain: partnerships, practices, and HIV risk behaviours', *The Lancet*, 358 (9296), pp. 1835-1842.

Joubert, G. & Ehrlich, R. (2007) *Epidemiology: a research manual for South Africa*. 2nd ed. Cape Town: Oxford University Press Southern Africa.

Maree, K. (2010) 'Statistical analysis II: inferential statistics'. In: *First steps in research*. Pretoria: Van Schaik.

McCarty, C. (2003) 'Differences in response rates using most recent versus final dispositions in telephone surveys', *Public Opinion Quarterly*, 67, pp. 396-406.

Mosavel, M., Simon, C., Van Stade, D. & Buchbinder, M. (2005) 'Community-based participatory research (CBPR) in South Africa: engaging multiple constituents to shape the research question', *Social Science & Medicine*, 61 (12), pp. 2577-2587.

Mouton, J. (2002) 'Data collection (sources of error)'. In: *Understanding social research*. Pretoria: Van Schaik.

O'Rourke, D. & Blair, J. (1983) 'Improving random respondent selection in telephone surveys', *Journal of Marketing Research*, XX, pp. 428-432.

Ohenjo, N., Wills, R., Jackson, D., Nettleton, C., Good, K. & Mugarura, B. (2006) 'Health of indigenous people in Africa', *The Lancet*, 367(9526), pp. 1937-1946.

Robins, J.M., Rotnitzky, A. & Scharfstein, D.O. (2000) 'Sensitivity analysis for selection bias and unmeasured confounding in missing data and causal inference models', *The IMA Volumes in Mathematics and its Applications*, 116, pp. 1-94.

Roth, T., Coulouvrat, C., Hajak, G., Lakoma, M.D., Sampson, N., Shahly, V., Shillington, A.C., Stephenson, J.J., Walsh, J.K. & Kessler, R.C. (2011) 'Prevalence and perceived health associated with insomnia based on DSM-IV-TR; international statistical classification of diseases and related health problems, tenth revision; and research diagnostic criteria/international classification of sleep disorders, second edition criteria: results from the America insomnia survey', *Biological Psychiatry*, 69, pp. 592-600.

Scott, A., Jeon, S., Joyce, C.M., Humphreys, J.S., Kalb, G., Witt, J. & Leahy, A. (2011) 'A randomized trial and economic evaluation of the effect of response mode on response rate, response bias, and item non-response in a survey of doctors', *BMC Medical Research Methodology*, 11, p. 126.

Sturgis, P., Smith, P. & Hughes, G. (2006) *A study of suitable methods for raising response rates in school surveys* [Online]. Department for Education and Skills. United Kingdom.

Thorpe, C., Ryan, B., MacLean, S.L, Stewart, M., Brown, J.B., Reid, G.J. & Harris, S. (2008) 'How to obtain excellent response rates when surveying physicians', *Family Practice*, 26 (1), pp. 65-68.

VanGeest, J.B., Johnson, T.P. & Welch, V.L. (2007) 'Methodologies for improving response rates in surveys of physicians: a systematic review', *Evaluating Health Professionals*, 30 (4), pp. 303-321.

Welman, C., Kruger, F. & Mitchell, B. (2005) 'Population and sampling types'. In: *Research methods.* 3rd ed. Cape Town: Oxford University Press.

Whittmore, R. & Knafl, K. (2005) 'The integrative review: updated methodology', *Journal of Advanced Nursing*, 52 (5), pp. 546-553.

Woolf, S.H., Rothemich, S.F., Johnson, R.E. & Marsland, D.W. (2000) 'Selection bias from requiring patients to give consent to examine data for health services research', *Archives of Family Medicine*, 9 (10), pp. 1111-1118.

Chapter 10

Cohort Studies

The discussions covered in this chapter will cover the following topics:

- A cohort study design
- Information analysis scenario of a cohort study versus case control study

10.1 A cohort study design

LET US FOCUS ON THE strengths and weaknesses of the choice of a cohort study design in addressing the question of whether alcohol consumption is associated with the onset of dementia. A cohort study is a type of epidemiologic study and is also known as a follow-up study or a longitudinal study of disease incidence, and the follow-up may be prospective or retrospective (Cromar et al, 2006).

It is considered as the main strength of cohort studies that they have the capacity to study change and development of outcome over a time period, which gives the researcher a measure of control over the variables being studied and to assess a range of outcomes associated with the exposure being studied (Saunders et al, 2003).The fact that there is case ascertainment in cohort studies is a strength because this ensures that there is better quality data collected, which means that it is very likely known with certainty that the exposure preceded the disease, considering the possible causal relationship between exposure and disease (Bruce et al, 2008). In this regard, the fact that the relative risk of onset of dementia as a result of alcohol consumption can be directly measured during the course of a cohort study can also be considered as the strength of the choice of a cohort study design (Joubert & Ehrlich, 2007). Unlike a prospective cohort study, it is a strength that a retrospective cohort study may be quicker to conduct as it makes use of historically completed data (Joubert & Ehrlich, 2007). Retrospective cohort studies are useful if good historical records are available about the exposure and outcome, otherwise, this could be a weakness.

A prospective cohort study's major weakness is that it may be time-consuming, demanding and expensive to follow-up cases to see if they develop dementia because some of the people in the cohort may be lost due to death, relocation or refusal to continue participating before the cohort study is concluded (Christensen, 2007). The observation by Walker & Shaper (1984) that it is a challenge to obtain accurate mortality data and more so, morbidity data, is a valid point because accurate medical record keeping is a major issue in the sub-Saharan region, and this could pose a major challenge when conducting cohort studies. A very large sample could be required if alcoholic dementia is rare in the population of interest (Joubert & Ehrlich, 2007). The fact that the level of alcohol consumption could change during the course of the cohort study, and that there could be a long time between exposure and outcome could be a potential source of random error and or bias (Bruce et al, 2008). It is interesting that some authors observe that a cohort study design is not representative of the population, because different samples of the population could yield different relative risks of the association between alcohol consumption and the onset of dementia (Welman et al, 2003).

The strengths and weaknesses of the choice of a cohort study to investigate the association between alcohol consumption and dementia indicate that it may be necessary for the researchers to justify the choice of the cohort study. Confounding factors should not be ignored when investigating the association between exposure and outcome (Wannamethee et al, 2001).

The researchers need to decide whether a prospective cohort study design or a retrospective study design would be the most appropriate to use in investigating whether alcohol consumption is associated with the onset of dementia. For instance, if the researchers have budget constraints and cannot afford to study or follow-up a very large sample, a retrospective cohort study design would be the most appropriate. A retrospective cohort study is usually less expensive than a prospective cohort study design and it shortens the time needed to conduct a cohort study as it makes use of historically compiled data (Cromar et al, 2006).

It may be useful for the researchers to have some prior idea of the prevalence of dementia, as it is advisable that a cohort study should be undertaken only when there is likely to be a sizeable incidence of disease (Joubert & Ehrlich, 2013). For example, literature indicates that dementia is rare below 55 years of age, 5-10% prevalent above 65 years of age, 20% prevalent above 80 years of age, and 70% prevalent above 100 years of age (Longmore et al, 2004). In this regard, if the researchers wish to employ a prospective cohort study design to follow a cohort of young men and women aged below 55 years of age, they may have to follow them for many years before they could be certain that there is association or no association between consumption of alcohol and the onset of dementia. A retrospective cohort study design may be the most appropriate if the researchers wish to investigate whether alcohol consumption is associated with the onset of dementia in a cohort of elderly men and women, but this approach could be useful if good records are available so that an association between alcohol consumption and onset of dementia could be investigated.

However, in the event of good records not being available, a prospective cohort study design may be appropriate. Suppose in 2022 the researchers recruit a total of 500 men and women aged below 55 years of age, in the Swaziland city of Mbabane, and divide them into exposure group and non-exposure group with regard to alcohol consumption and follow them up for a period of 20 years. Suppose during this 20 year period, 10 new cases of dementia are diagnosed in the group that consumed alcohol daily and 20 new cases are diagnosed in the group that did not consume alcohol daily. According to Bruce et al (2008), the relative risk can be calculated as the incidence of disease in exposed group divided by the incidence of disease in unexposed group. In this case, the relative risk of the onset of dementia among those who consumed alcohol daily is 0.5 times the risk of the onset of dementia in those who did not consume alcohol daily. If these results are real and $p<0.05$, the researchers may conclude that consuming alcohol daily has protective effects from the onset of dementia, and the null hypothesis cannot be rejected.

According to (Cromar et al, 2006), retrospective cohort studies are particularly useful for unusual exposures or occupational exposures. Alcohol consumption is not really an unusual exposure, and a prospective cohort study design would be more appropriate in the investigation of whether alcohol consumption is associated with the onset of dementia.

Another point that is worth highlighting is the selection of subjects for the cohort study. Cromar et al (2006), mention that for common risks such as alcohol consumption, investigators may enrol a general population cohort, and for uncommon risks such as occupational exposures to hazardous chemicals, investigators use special exposure cohorts. Since alcohol consumption is classified as a common risk factor, the researchers should enrol a particular subset of the general population which may be easy to follow up.

Some researchers highlight that one of the weaknesses of prospective cohort study designs is that subjects may be lost to follow-up, which could result in bias. Loss to follow-up of participants is regarded as one of the major challenges or weaknesses of prospective cohort studies (Joubert & Ehrlich, 2007). Interestingly, through follow-up examinations in 1993-1994 and in 1997-1999, and an extensive monitoring system, nearly complete follow up of nearly 100 per cent was obtained in the Rotterdam Study (Ruitenberg et al, 2002). The Rotterdam Study commenced in 1990 and it employed a population-based prospective cohort study design to examine the relation between alcohol consumption and risk of dementia in 7983 individuals aged 55 years and older, from a suburb of Rotterdam (Ruitenberg et al 2002).

Much as researchers may wish to use a cohort study design to investigate whether alcohol consumption is associated with the onset of dementia. I still wonder if the diagnosis of dementia would not be problematic, as some authors argue that there are no proper criteria to determine for sure whether the dementia is as a result of a specific exposure or it may be due to other causes such as aging, or in fact due to an overlap of a number of factors (Reitz et al, 2007). The issue of determining whether there is a causal relationship between exposure and outcome is regarded as critical, and therefore it is important to use pointers such as the Bradford-Hill guidelines that can help researchers reach a conclusion in this regard (Bruce et al, 2008).

Cohort studies may require a large sample of people or cohort, and, a long follow up time, in order to collect enough information, which should ideally yield reliable results. According to Joubert & Ehrlich (2007), the reality is that an even larger sample may be required especially if the exposure of interest is rare, and sometimes an unnecessarily larger or smaller sample size may be estimated if such a study is still new and there is no published literature about the subject.

Since research to investigate the association of alcohol consumption is not really new, the researchers should do some background work or thoroughly search published literature on the subject to ensure that they make an informed decision about the sample size. The decision on sample size is critical, and it is strongly recommended that a statistician's advice be sought before embarking on the study (Bruce et al, 2008).

Mukamal et al (2003) are interesting when they report that it is very hard to find any previous cohort studies that have addressed the risk of confirmed dementia in a large cohort of adults with repeated measures of alcohol use. It seems as if there is an issue about confirming the diagnosis of dementia especially alcoholic dementia in cohort studies. It is also interesting to note that previous studies of alcohol consumption and dementia have reported conflicting results. In fact, some researchers report that there is no evidence of any association between alcohol consumption and the onset of dementia (Ruitenberg et al, 2002). On the other hand, some researchers report that there is an association between alcohol consumption and the onset of dementia, but what is not clear is the mechanism by which alcohol consumption could cause the onset of dementia (Hulse et al, 2005). It is possible that the conflicting results of the previous studies are due to bias.

While still on the issue of bias, I notice that some researchers observe that prospective cohort study designs are less vulnerable to bias because of the large samples, matching of subjects and adjustment of results. This is a valid point, but one wonders why there is such a growing evidence of conflicting results regarding similar cohort studies aimed at investigating whether there is an association between alcohol consumption and the onset of dementia.

Conducting a prospective cohort study to investigate whether there is an association between alcohol consumption and the onset of dementia, has its pros and cons. Of critical importance is to ensure that an informed decision is taken with regard to the sample size, and to eliminate or prevent bias.

10.2 Information analysis scenario of a cohort study versus case control study

- Null hypothesis: There is no association between taking aspirin daily and the onset of cancer
- The advantages of using a cohort study design compared to a case-control study design are:

- Case ascertainment in a cohort study design makes it known with certainty whether the exposure such as taking aspirin daily preceded the onset of disease such as cancer or not (Bruce et al, 2008). Whereas in a case-control study design, it is difficult to know whether the exposure preceded the disease or not because exposure status is determined after the onset of disease has occurred (Joubert & Ehrlich, 2007).
- The risk of developing a disease such as cancer can be directly measured in a cohort study design (Bruce et al, 2008). Whereas in a case-control study design the risk of exposure is usually restricted to the odds ratio and risk differences cannot usually be calculated. Moreover, information-recall bias may be a problem in case-control studies because the cases may not really remember all the exposures that they were exposed to especially if the exposures happened many years ago (Joubert & Ehrlich, 2007).
- A cohort study design enables the assessment of a range of diseases associated with the exposure such as taking aspirin daily being studied (Bruce et al, 2008). Whereas in a case-control study a range of exposures may be associated with the single outcome such as cancer (Joubert & Ehrlich, 2007).

The disadvantages of using a cohort study design compared to a case-control study design are:

- A prospective cohort study design may be time-consuming and expensive as it may require following-up on the selected group of elderly men and elderly women being surveyed for an extended period especially if the cancers occur a long time after the exposure (Cromar et al, 2006). Whereas, in a case-control study, there is no need to follow-up on the selected cohort because cases are selected when the disease has already occurred (Joubert & Ehrlich, 2007).
- A cohort study design may require a very large sample of elderly men and elderly women to be surveyed especially if the cancers are very rare. Whereas a case-control study design, does not require a larger sample, and may be considered as an efficient form of a cohort study (Joubert & Ehrlich, 2007).
- A cohort study may be prone to bias from loss of cases or inconsistent or poor follow-up especially if the cohort needs to be followed for an extended period (Christensen, 2007). Whereas, a case-control study is less prone to such bias because it does not require extended follow-up of cases as the disease has already occurred (Joubert & Ehrlich, 2007).

- A sampling frame is a complete list of the units of analysis in which each unit of analysis is mentioned only once. For example, a clinic patients' register (Welman et al, 2005).
- Case ascertainment involves case definition to ensure that a standard criteria is used define the specific disease of focus in the study. Case ascertainment also involves case finding to ensure with certainty that the right people are included in the study (Bruce et al, 2008).
- According to Bruce et al (2008) relative risk (RR) can be calculated as follows:

Relative risk = Incidence of disease in exposed group / Incidence of disease in non- exposed group

Relative risk = 40 / 141

Relative risk = 0.28

Therefore, the relative risk of colon cancer for those who took aspirin daily was 0.28 times that of those who did not take aspirin daily. Since the relative risk is less than 1, this implies that taking aspirin daily had protective effect from developing colon cancer.

- The relative risk of 1.00 implies that the risk of kidney cancer among the elderly men and elderly women who did not take aspirin daily was the same. The relative risk of 6.28 indicates that there was a 6.28 times risk of developing kidney cancer among men who took aspirin daily compared to those who did not. The $p<0.05$ indicates that the risk of developing kidney cancer was significant, and therefore the null hypothesis can be rejected.

The relative risk of 1.00 implies that the risk of cancer among women and men who did not take aspirin daily was the same. The relative risk of 2.11 indicates that there was a 2.11 risk of developing kidney cancer among the elderly women who took aspirin daily compared to those who did not.

REFERENCES

Bruce, N., Pope, D. & Stanistreet, D. (2008) 'Cohort studies'. In: *Quantitative research methods for health research: a practical guide to epidemiology*. Chichester: John Wiley & Sons, Ltd

Christensen, L.B. (2007) 'Reliability and validity in experimental research'. In: *Experimental methodology*. Boston: Pearson

Cromar, N., Cameron, S. & Fallowfield, H. (2006) 'Environmental epidemiology'. In: *Environmental health in Australia and New Zealand*. Melbourne: Oxford University Press.

Hulse, G.K., Lautenschlager, N.T., Tait, R. & Almeda, O.P. (2005) 'Dementia associated with alcohol and other drug use', *International Psychogeriatrics*, 17, pp. S109-S127.

Joubert, G. & Ehrlich, R. (2007) 'Study design'. In: *Epidemiology: a research manual for South Africa*. 2nd ed. Cape Town: Oxford University Press Southern Africa.

Longmore, M., Wilkinson, M. & Rajagopalan, S. (2004) *Oxford handbook of clinical medicine*. New York: Oxford University Press.

Mukamal, K.J., Kuller, L.H., Fitzpatrick, A.L., Longstreth, W.T., Mittleman, M.A. & Siscovick, D.S. (2003) 'Prospective study of alcohol consumption and risk of dementia in older adults', *Journal of the American Medical Association*, 289 (11), pp. 1405-1413.

Reitz, C., den Heijer, T., van Duijn, C., Hofman, A. & Breteler, M.M.B. (2007) 'Relationship between smoking and risk of dementia and Alzheimer disease', *Neurology*, 69, pp. 998-1005.

Ruitenberg, A., van Swieten, J.C., Wittenman, J.C.M., Mehta, K.M., van Duijn, M., Hofman, A. & Breteler, M.M. (2002) 'Alcohol consumption and risk of dementia: the Rotterdam study', Lancet, 359, pp. 281-286.

Saunders, M, Lewis, P. & Thornhill, A. (2003) 'Deciding on the research approach and choosing a research strategy'. In: *Research methods for business students*. Harlow: Prentice Hall

Walker, M. & Shaper, A.G. (1984) 'Follow up of subjects in prospective studies based in general practice', *Journal of the Royal College of General Practitioners*, 34 (264), pp. 365-370 [Online].

Wannamethee, S.G., Shaper, A.G. & Walker, M. (2001) 'Physical activity and risk of cancer in middle-aged men', *British Journal of Cancer*, 85 (9), pp. 1311-1316.

Welman, C., Kruger, F. & Mitchell, B. (2005) 'Types of quantitative research designs'. In: *Research methodology*. Cape Town: Oxford University Press.

Chapter 11

Case-Control Studies

This chapter covers the following topics:

- Desirability and appropriateness of case-control studies
- Obtaining a suitable study population of cases and controls
- Bias and potential limitations of case-control studies
- Analysis of case-control data

11.1 Desirability and appropriateness of case-control studies

IN VITRO STUDIES HAVE shown that small amounts of aspirin may inhibit cyclogenase dependent platelet aggregation, suggesting that it may have clinically useful antithrombotic effect *in vivo* (Peto et al, 1988). Results that have been obtained *in vitro* may not necessarily be duplicated *in vivo* because there are many factors that may influence the outcome *in vivo*, such as the issue of non-adherence to treatment, among others. Suppose we would like to conduct a study to identify implications of non-adherence for a double-blinded randomized controlled trial to investigate how aspirin reduces the risk of stroke in middle-aged male doctors.

According to Joubert & Ehrlich (2007), randomized controlled trials, in which the research team randomly allocates eligible people to receive or not to receive one or more interventions that are being compared, are regarded as the most rigorous experimental designs in epidemiology. Double-blinded randomized controlled trials, are a type of randomized controlled trials in which both the subjects and the researchers or staff carrying out the treatments do not know to which trial group an individual has been assigned (Bruce et al, 2008).

Non-adherence could be minimized at the design stage of the randomized controlled trials by ensuring that a large sample of willing, motivated and informed subjects is recruited, and that a thorough baseline assessment is conducted with a clear inclusion and exclusion criteria and a clear follow-up plan (Bollinger et al, 2000). The recruitment of doctors in this study is important because doctors are very likely to understand the importance of adherence and the implications of non-adherence of such interventions, and doctors are also very likely to report accurate or reliable information with regard to illnesses that they may suffer (Peto et al, 1988). This is important because doctors who suffer from illnesses such as peptic ulcers or whose health condition(s) could be severely aggravated by participating in this study or who have very busy schedules that could negatively affect their participation in this study, could be excluded at the design stage of the study.

The implications of non-adherence to the study intervention are that the real effect of the intervention could either be overestimated or underestimated. If there is non-adherence, and those who do not adhere are excluded when the results are analysed, the results may show that either the intervention group or the control group had a marked advantage over the other, and this could lead to bias (Bruce et al, 2008). Drawing conclusions based on the exclusion of non-adherers could be misleading because the real effect of the aspirin could be overestimated.

On the other hand, if those who have not adhered are included in their original groups when the results are analysed, the non-adherers could result in an underestimation or dilution of the real effect of the intervention or aspirin in reducing the risk of stroke in middle-aged male doctors. At least, the inclusion of non-adheres during the analysis gives a realistic picture of what could happen in practice, as we know that some patients may adhere and some may not adhere with the intervention or treatment (Bruce et al, 2008). Perhaps, another double-blinded controlled trial should also look at females to see if there are any gender specific issues of adherence to the intervention.

11.2 Obtaining a suitable study population of cases and controls

ALTHOUGH STROKE IS uncommon below the age of 50, it is reported to be a prominent cause of disability especially among middle-aged males since they are considered to be at a 1.5 times higher risk of suffering a stroke than females (Edwards et al, 1996). In that sense, and based on their academic training and clinical practice experience, it would be expected that middle-aged male doctors to know better that they are at a higher risk of suffering a stroke especially if there is elevated blood pressure. Since elevated blood pressure is linked with the risk of stroke, it is interesting to note that elevated blood pressure tends to rise with age and there is corresponding rise in the prevalence of hypertension (Timmis & Nathan, 1994).

However, the risk is reduced even in middle-aged men who are healthy and who lead a healthy lifestyle. Since it is often said that prevention is better than cure, it would be expected that the middle-aged doctors to be keen on taking any prophylactic measures or treatment to ensure that they reduce the risk of suffering a stroke even if it is by any small amount. Of course the middle-aged doctors would want to know if taking such prophylactic measures or treatment outweighs any accompanying side effects. Peto et al (1988) observes that to answer such questions, direct evidence is needed from randomised controlled trials. One would therefore, expect the middle-aged male doctors to willingly take part and fully comply with this double-blinded randomized controlled trial aimed at investigating how aspirin reduces the risk of stroke in middle-aged male doctors. Various studies highlight the importance of ensuring that willing and motivated subjects are selected to participate in such research studies (Bruce et al, 2008). There is emphasis on the importance of minimizing non-adherence at the design stage.

It is interesting to note that some researchers observe that since there have not been many studies devoted to adherence, it cannot be overlooked that there is a likelihood of intentional non-adherence especially if the side effects of the intervention drug are very serious and unbearable over an extended period (Kim et al, 2007). One would assume that the middle-aged male doctors have had enough clinical experience to come up with different strategies of taking the drug in question in order to minimize its side effects. Walker et al (2006) report that adherence is increased if the intervention is preventative compared to curative, and even if there are side effects, the odds of adherence or compliance are increased as participants report multiple strategies to take medication.

The fact that the subjects in this study are middle-aged male doctors is very likely to increase the odds of adherence. According to Peto et al (1988), doctors are thought to be particularly suitable for these trials because of their particular ability to appreciate what is entailed and to judge the potential risks and benefits that may be associated with the prophylactic use of aspirin.

11.3 Bias and potential limitations of case-control studies

BLINDING IS REGARDED as one of the strengths of the ideal randomized controlled trial because it reduces the likelihood of bias (Bruce et al, 2008). In the context of a double-blinded randomized controlled trial, blinding is important because it reduces the likelihood of bias owing to differences in perceived response to treatment, sometimes called performance or reporting bias on the part of participants, and it also prevents assessment, diagnostic or detection bias on the part of healthcare providers or researchers (Joubert & Ehrlich, 2007).

The fact that both the subjects and the researchers are blinded in the double-blinded randomized controlled trial aimed at investigating how aspirin reduces the risk of stroke in middle-aged male doctors is particularly significant in this study. If the subjects (doctors) in this study are not blinded, they may already have an idea about the pharmacology of the drug and as a result they may have perceived response to the treatment, including the expected therapeutic effects as well as the expected side effects of the treatment. If the subjects perceive that the drug has serious side effects, they may not adhere or they may end up withdrawing from the study. According to Longmore et al (2004), the effects of aspirin are dose-related and potentially fatal. Those subjects taking the placebo may also not adhere because they may already have perceived that the placebo has no therapeutic effect. The issue here is that in addition to there being a reporting bias, such non-adherence could potentially result in underestimation of the effect of the intervention. On the other hand, if the subjects taking aspirin are excited about the perceived prophylactic effects of aspirin, they may very keenly adhere and report positively about the intervention (Peto et al, 1988). Such adherence and positive self-reporting by the subjects (doctors) could potentially overestimate the therapeutic effects of the drug (Bruce et al, 2008).

If the researchers or investigators are not blinded, bias may also creep in because the researchers may end up paying more attention to those subjects receiving the drug or aspirin and overlook those that are receiving the placebo or control. Those taking the placebo may end up not adhering, or they may withdraw from the study because of poor follow up especially if the study is challenging and long (Day et al, 2002). This could therefore result in assessment or detection bias on the part of the researchers. At the analysis stage, there could be a potential overestimation of the therapeutic effect of the drug or aspirin in reducing the risk of stroke (Bruce et al, 2008).

Blinding of the subjects and the researchers is significant in the double-blinded randomized controlled trial aimed at investigating how aspirin reduces the risk of stroke in middle-aged male doctors. This is particularly so, because such blinding eliminates performance or reporting bias on the part of the subjects, and also eliminates assessment or detection bias on the part of the researchers (Joubert & Ehrlich, 2007).

Physicians would be preferred as the target population when conducting research studies, such as double-blinded randomized controlled trials to investigate how aspirin reduces the risk of stroke, because physicians are very likely to report accurately about their health status compared to members of the general public. It is reported that the characteristics of those who are recruited to participate in randomized control trials differ from those who are not, albeit inconsistently, in ways that potentially affect outcome (Patterson et al, 2010).

This suggests that the results obtained from the double-blinded randomized controlled trial to investigate how aspirin reduces the risk of stroke in middle-aged doctors, cannot be generalized or not considered to be representative of all middle-aged men. There is a possibility that even if the results of the double-blinded randomized controlled trial demonstrate clear benefits for middle-aged doctors, the proportion of eligible middle-aged male patients who may qualify for treatment from the general public may be low due to the variable-risk benefit, raising concerns about application to general medical practice (Evans & Kalra, 2001).

Such a study may not only be non-representative of all middle-aged men in the general public, but it may also be non-representative of all middle-aged doctors, because settings may also differ. For example, results obtained from a double-blinded randomized controlled trial conducted among middle-aged doctors in a certain part of America may yield vastly different results from a similar study conducted in a certain part of Africa. Perhaps, that is why it would make more sense if large studies of how aspirin reduces the risk of stroke among middle-aged doctors were conducted across different parts of the world, to at least come up with some semblance of generalizability among middle-aged doctors, if at all (Peto et al, 1988). According to Wilson et al (2000), trials in any setting are rarely fully representative with respect to both patient and disease related characteristics. The unfortunate danger is that the evidence derived from randomized controlled trials is usually considered sufficient enough to base the general public's treatment protocols on. It is hardly surprising that some researchers argue if randomized controlled trials are really the only gold standard that glitters (Slade & Priebe, 2001).

Much as randomized controlled trials are regarded as the gold standard in terms of evidence based research, the challenge is that the outcomes cannot always be inferred to be representative of the general public. The results obtained from the double-blinded randomized controlled trial to investigate how aspirin reduces the risk of stroke in middle-aged doctors, cannot be regarded to be representative of all middle-aged men in the general public, due to the variable-risk benefit.

11.5 Analysis of case-control data

- Researchers may choose 2 controls per case because they might have found it difficult to decide which approach would deliver the most appropriate controls (Bruce et al, 2008). By choosing 2 controls per case, the researchers might have also wanted to increase the statistical power of the case study (Bruce et al, 2008).

- The researchers could match controls for month and year of birth and maternity unit of birth because they might have wanted to overcome confounding by removing the possibility of studying the association between a confounding factor and the outcome (Bruce et al, 2008). This also avoids the implication of having to deal with unmatched analysis which is less efficient than matched analysis.
- The researchers asked about parental smoking history and birth weight because parental smoking is regarded as one of the risk factors for sudden infant death syndrome, and they also wanted to see whether the cases might have been more exposed to the risk factors than the controls

Variable	Cases (n=146)	Controls (n=275)	Odds ratio	95% confidence interval
Father smokes	11	45	1.72	0.94-3.13
Child sleeps on side (compared to supine)	75	104	1.74	1.16-2.61
Sex of child male (Compared to female)	105	108		2.56-6.12

Odds ratio = Odds of being exposed if a case / Odds of being exposed if a control

Odds ratio for being a male compared to being a female child = 105/108 = 0.97

The odds ratio is less than 1, and this indicates that there is no association between sex of child and the risk of SIDS. However, the odds ratio falls outside the 95% confidence interval (2.56-6.12), and this 95% confidence interval is very wide, which indicate that the estimate is very imprecise.

- The odds ratio is 1.72 and is greater than 1, which indicates that the risk of SIDS is greater among children who are exposed to their fathers who smoke. However, the wide confidence interval (0.94-3.13) could imply that the estimate is imprecise (Joubert & Ehrlich, 2007).

REFERENCES

Bollinger, C.T., Zellweger, J., Danielsson, T., van Biljon, X., Robidou, A., Westin, A., Perruchoud, A.P. & Sawe, U. (2000) 'Smoking reduction with oral nicotine inhalers: double blind, randomized clinical trial of efficacy and safety', *British Medical Journal,* 321, pp. 329-333.

Bruce, N., Pope, D. & Stanistreet, D. (2008) 'Intervention studies'. In: *Quantitative research methods for health research: a practical guide to epidemiology*. Chichester: John Wiley & Sons, Ltd.

Day, L., Fildes, B., Gordon, I., Fitzharris, M., Flamer, H. & Lord, S. (2002) 'Randomised factorial trial of falls prevention among older people living in their own homes', *British Medical Journal,* 325, pp. 1-6.

Edwards, C.R.W., Bouchier, I.A.D., Haslett, C. & Chilvers, E.R. (1996) *Davidson's principles and practice of medicine*. Edinburgh: Churchill Livingstone.

Evans, A. & Kalra, L. (2001) 'Are the results of randomized controlled trials on anticoagulation in patients with atrial fibrillation generalizable to clinical practice?', *Archives of Internal Medicine,* 161, pp. 1443-1447.

Joubert, G. & Ehrlich, R. (2007) 'Study design'. In: *Epidemiology: a research manual for South Africa.* 2nd ed. Cape Town: Oxford University Press Southern Africa.

Kim, E.Y., Han, H.R., Jeong, S., Kim, K.B., Park, H., Kang, E., Shin, H.S., Kim, M.T. (2007) 'Does knowledge matter?: International medication nonadherence among middle-aged Korean Americans with high blood pressure, *The Journal of Cardiovascular Nursing,* 22 (5), pp. 397-404.

Longmore, M., Wilkinson, M. & Rajagopalan, S. (2004) *Oxford handbook of clinical medicine*. New York: Oxford University Press.

Patterson, S., Kramo, K., Soteriou, T. & Crawford, M.J. (2010) 'The great divide: a qualitative investigation of factors influencing researcher access to potential randomized controlled trial participants in mental health settings', *Journal of Mental Health*, 19 (6), 532-541.

Peto, R., Gray, R., Collins, R., Wheatley, K., Hennekens, C., Jamrozik, K., Warlow, C., Hafner, B., Thompson, E., Norton, S., Gilliland, J. & Doll, R. (1988) 'Randomised trial of prophylactic daily aspirin in British male doctors', *British Medical Journal*, 296, pp. 313-316.

Slade, M. & Priebe, S. (2001) 'Are randomized controlled trials the only gold that glitters?' *The British Journal of Psychiatry*, 179, pp. 286-287.

Timmis, A.D. & Nathan, A. (1994) *Essential of cardiology*. London: Blackwell Scientific Publications

Walker, E.A., Molitch, M., Kramer, M.K., Kahn, S., Ma, Y., Edelstein, S., Smith, K., Johnson, M.K., Kidabchi, A. & Crandall, J. (2006) 'Adherence to preventive medications', *Diabetes Care*, 29, pp. 1997-2002

Wilson, S., Delaney, B.C., Roalfe, A., Roberts, L., Redman, V., Wearn, A.M. & Hobbs, F.D.R. (2000) 'Randomised controlled trials in primary care: case study', *The British Medical Journal*, 321, pp. 24-27

Chapter 12

Critical Appraisal of Research Evidence

This chapter covers the following topics:

- Finding research evidence
- Appraising research evidence
- Critical analysis

12.1 Finding research evidence

DEATH CERTIFICATE DATA can be of value for the purposes of epidemiological research, as epidemiological research helps us to practically make associations between information or data and risk factors for disease outcomes in groups or populations. Joubert & Ehrlich (2005) acknowledge that the notion of cause or risk factor could be tricky because most conditions require a number of causes or risk factors working together before the disease outcome could occur. It is of paramount importance to realize that the validity and reliability of each data element has to be taken into consideration (Bruce et al, 2008).

According to Cohen et al (2007), death certificates contain important information such as age, sex, race, occupation, residential address, date and place of death which could be used in epidemiological studies for the purpose of investigating possible factors associated with breast cancer. The approach would be to identify a country and analyse the death certificate data for a specific period such as the year 2000 to the year 2010, and then separate the information by age, sex, occupation, residential address, and place of death. One would conduct a case-control study, and the cases would be all deaths due to breast cancer and my controls would be all deaths other than cancer.

The strengths of death certificate data include its relative ready availability and relative low cost which would make it quicker to interpret the data; and its completeness would allow for interpretation and description of patterns of in-country and cross-country populations (Cohen et al, 2007). The other strength is that death certificate data can be used as a baseline, particularly in states or countries where the use of mammography is not yet well established (Geffken, 2000). Since all types of death are reported on death certificates, it implies that the interpretation of the data would have a lesser probability of being biased.

The limitations of death certificate data include the inconsistencies with which the variables are recorded in different states or countries (Geffken, 2000). Such variations would affect the interpretation of the data in the sense that one would not be able to confidently apply possible factors associated with breast cancer from one country to another country. One would also consider it as a weakness the fact that the recording of the cause of death is largely dependent on the physician, because sometimes the physician would just record it as a medical condition without being specific. The recording of the cause of death as being a medical condition has somewhat become a norm, particularly in South Africa and the Kingdom of Swaziland. This implies that using death certificate data to investigate possible factors associated with breast cancer could be a challenge. Hoel et al (1993) highlights the challenge posed by cancer mortality studies because of the issues of quality of the data.

Death certificate data presents a good opportunity to investigate possible factors associated with breast cancer. However, the strengths and limitations need serious consideration as these could have health policy implications.

12.2 Appraising research evidence

1. The published paper to be critically appraised focused on evaluating the use of a rehabilitation service to prevent falls in the community for older people who had called an ambulance after a fall but had not been taken to hospital (Logan et al, 2010).
2. Inclusion criteria for this study:
    a. A person had to be aged over 60, and
    b. Lived at home or in a care home in one of four primary care trust areas (Nottingham City, Rushcliffe, Broxtowe and Hucknall, and Gedling) in Nottinghamshire, United Kingdom, and
    c. Had contacted the East Midlands Ambulance Service through the emergency telephone system because of a fall but had not been taken to hospital.
3. Exclusion criteria:
    a. If the person was unable to give consent,
    b. If the person was deemed too ill to participate,
    c. If the person was already in a falls prevention rehabilitation programme.
4. Randomization
    a. The groups were successfully randomized through the use of a computer generated randomization scheme, and participants had equal chance of being assigned to the intervention group and the control group (Logan et al, 2010). The median (interquartile range) age (years) were almost similar between the intervention group [82 (78-86)] and the control group [83 (77-86)]. The proportion of

men in the intervention group [37 (36)] was almost similar to the proportion of men in the control group [35 (34)], and the proportion of women in the intervention group [65 (64)] was also almost similar to the proportion of women in the control group [67 (66)].

   b. If any factors are not balanced, researchers should arrange a comparison between the intervention and control group in a way that avoids the influence of confounding factors (Bruce et al, 2008). Researchers should factor for or adjust for the influence of the factors that are not balanced (Peto et al, 1988).

5. Blinding
   a. Blinding means that the participants and study staff are prevented from knowing which treatment a subject is receiving in the randomized controlled trial (Christensen, 2007).
   b. This trial was blinded in the following ways: The assessors who contacted the participants and collected missing data on outcomes were blinded and the research staff that captured data were also blinded to the allocation groups.
6. The main outcome measure in this study was "rate of falls per person per year".
7. Since the outcome of rates of falls per person per year has an odds ratio of 0.45, with a confidence interval of 0.35 to 0.58, it implies that the chances of a person in the intervention group falling per year is 45 per cent less or 0.45 times the risk of a person in the control group. It also shows that there is 95 per cent confidence that the odds ratio falls within the interval 0.35 to 0.58.
8. The researchers in this study carried out an intention to treat

analysis. "Intention to treat analysis" means that the subjects were analysed according to the groups in which they were initially allocated, regardless of the effect of intervention received (Joubert & Ehrlich, 2007).

9. Given the fact that the number of diaries returned between the two groups did not differ significantly (P=0.36, Mann-Whitney U test):

- We would want to know if there was a difference in return rates between the two groups because we would want to know if the difference was significant or not, and the degree to which the difference was significant or not, so that we can tell if there was any effect of the intervention or not, and the degree of the effect or no effect.
- Researchers used a hypothesis test because they wanted to quantify the degree to which chance variability might explain any observed differences between the two groups, and the Mann-Whiney U test was used in particular because the data did not meet the assumptions of parametric hypothesis tests (Bruce et al, 2008).
- The p value of 0.36 tells us that the difference between the intervention and the control was not significant since $p>0.05$

1. Two types of bias that could have occurred in this study are: selection bias and performance bias. Researchers in this study tried to reduce bias by concealment of the allocation process and by blinding the assessors and research staff who collected and captured the data (Logan et al, 2010).

12.3 Critical analysis

THE EPIDEMIOLOGY OF mycobacterium tuberculosis and the use of secondary data sources

Summary:

Mycobacterium tuberculosis continues to be a huge public health issue across the globe, and it is estimated that approximately 8.7 million new cases of mycobacterium tuberculosis were recorded in the year 2011. The seriousness and magnitude of the HIV and mycobacterium tuberculosis co-infection seems to undermine whatever efforts have been made to deal with this challenge especially in resource-strained developing countries such as South Africa. There is indication that all countries need to double their efforts and ensure that they strengthen their health systems to enable a sustainable tuberculosis care and control. The main source of data for this report was the Global Tuberculosis Report 2012. It has to be acknowledged that secondary sources of data have their strengths and limitations, and the Global Tuberculosis Report is no exception. The main strengths of secondary data sources are that they save time and are cost-effective because the data is already there. Secondary data obtained from an external source is likely to be objective. The main limitations of secondary sources of data are that the consumer of the secondary data has no control of what has already been captured especially if there are issues about the quality and validity of the data. Moreover, the secondary data may have been collected by other people for different purposes which may lack specificity and not be compatible with what one would like to use the secondary data for.

Background and Epidemiology of mycobacterium tuberculosis:

Mycobacterium tuberculosis is a notifiable infectious disease that typically affects the lungs but can also spread to other parts of the body as well, and it is reported to be second to HIV as the leading cause of death from an infectious disease worldwide (World Health Organization, 2012). Although it is encouraging to see that all six regions of the World Health Organization have shown a decrease in the mortality and incidence rates of mycobacterium tuberculosis, the global burden of mycobacterium tuberculosis is very worrying as there was an estimated 8.7 million new cases of mycobacterium tuberculosis in 2011, of which 13 per cent were co-infected with HIV (World Health Organization, 2012). The risk posed by mycobacterium tuberculosis, including multi-drug resistant tuberculosis (MDR-TB) and extensively drug resistant tuberculosis (XDR-TB) is enormous, and this requires each and every country to have systems in place and to scale up their efforts in order to effectively deal with this enormous challenge, as weak health systems have been reported to be the main reason why tuberculosis care and control are being hindered (Atun et al, 2010).

Much as the heaviest burden of mycobacterium tuberculosis is carried by the developing countries, it is interesting to note that even some of the developed countries are facing a huge challenge in this regard. For example, unlike in most developed countries, rates of MDR-TB continue to rise in the United Kingdom, and this should be a wake-up call to all developed countries, and this wake-up call should remind all of us that more sustained efforts are required if we are to win the battle against *mycobacterium tuberculosis* (Zumla et al, 2013).

Mycobacterium tuberculosis ranks as one of the foremost infectious diseases in developing countries, and it is the cause of considerable morbidity and mortality especially among children (Nzimande, 1996). It is really a concern that the burden is highest in Asia and Africa, and that the African region has 24 per cent of the world's cases and the highest rates of cases and deaths per capita, and that South Africa is among the top five worst affected countries in the world (World Health

Organization). Along with HIV and AIDS which is causing havoc in the developing world, tuberculosis forms the biggest health problem in South Africa (de Haan, 2009). South Africa has recently been faced with an outbreak of multi-drug resistant tuberculosis (MRD-TB) and extensively-drug resistant tuberculosis (XDR-TB) which seem to be fuelled by the high incidence of HIV, as South Africa is considered to have the highest incidence of HIV in the world and the fifth highest incidence of mycobacterium tuberculosis and HIV co-infection in the world (World Health Organization, 2012). The dual havoc of HIV and tuberculosis undermines HIV and AIDS control efforts and tuberculosis control efforts in this resource-strapped country (Dikeledi et al, 2009). Although there has been some acknowledgement about the possible association between the high burden of mycobacterium tuberculosis and socio-economic status (Lonnroth et al, 2010), there is paucity of epidemiological research to confirm the exact pathways of this supposed relationship (Harling et al, 2008).

In South Africa, the problem of tuberculosis has severely impacted on the occupational setting especially in the mines, correctional services, rehabilitation centres, and healthcare facilities, most of which are overcrowded, and new cases continue to be diagnosed in foundry, ceramic, engineering and construction workers (Nogueira, 2009). The mining industry has been regarded as the hardest hit sector as studies have shown that miners and former miners have been exposed to silica dust which can cause silicosis (Hattingh & Acutt, 2003). It is reported that there is an increased incidence of mycobacterial disease in silicosis especially where background tuberculosis and HIV rates are high, as in South Africa (LaDou, 2007). The negative impact that mycobacterium tuberculosis is causing in the occupational setting is very profound because this results in increased absenteeism, decreased productivity, and increased medical and healthcare liability in many affected countries.

Discussion of data source:

The main secondary data source used for this paper was the Global Tuberculosis Report 2012. One has also used data from other sources such as the South African Demographic and Health Survey and related research studies as highlighted below. Each country supplies the World Health Reference Centre with information regarding cases using an online reporting system which automatically checks for any inaccuracies in the supplied data, and experts from the World Health Organization's headquarters or regional offices also follow up with each country in order to verify the information (World Health Organization, 2012). Since the Global Tuberculosis Report was produced in September 2012, there is a possibility that there might be discrepancies because it is possible that some countries might have submitted their data after this date.

Table 1: Reporting of data in the 2012 round of global TB data collection

	COUNTRIES AND TERRITORIES		MEMBER STATES	
WHO REGION OR SET OF COUNTRIES	NUMBER	NUMBER THAT REPORTED DATA	NUMBER	NUMBER THAT REPORTED DATA
African Region	46	46	46	46
Eastern Mediterranean Region	23	23	22	22
European Region	54	42	53	41
Region of the Americas	46	46	35	35
South-East Asia Region	11	11	11	11
Western Pacific Region	36	36	27	27
High-burden countries	22	22	22	22
WORLD	**216**	**204**	**194**	**182**

It can be noted from Table 1 that the European Region had the highest number of countries that did not report data, and this may have been due to the fact that these were low-incidence countries (World Health Organization, 2012).

Out of the 6.2 million mycobacterium notifications worldwide, the Africa region accounted for 24% of the cases in the year 2011 (World Health Organization, 2012). Interestingly South Africa had a notification of 993 per 100 000 population in 2011 (World Health Organization, 2012), compared to 948 per 100 000 population in the year 2007 (Dheda et al, 2010). It is also reported that South Africa had an estimated incidence of 600 per 100 000 population in 2005 (Harling et al, 2008) and an incidence of 218 per 100 000 population in 2003 (den Boon, 2007). The abovementioned trends show an obvious continued increase in the incidence of mycobacterium tuberculosis in South Africa. According to Joubert & Ehrlich (2007), Cape Town has consistently been among the cities with the highest incidence of mycobacterium tuberculosis in South Africa during the past two decades.

Figure 1: Estimated mycobacterium incidence rates for South Africa, 2003-2011

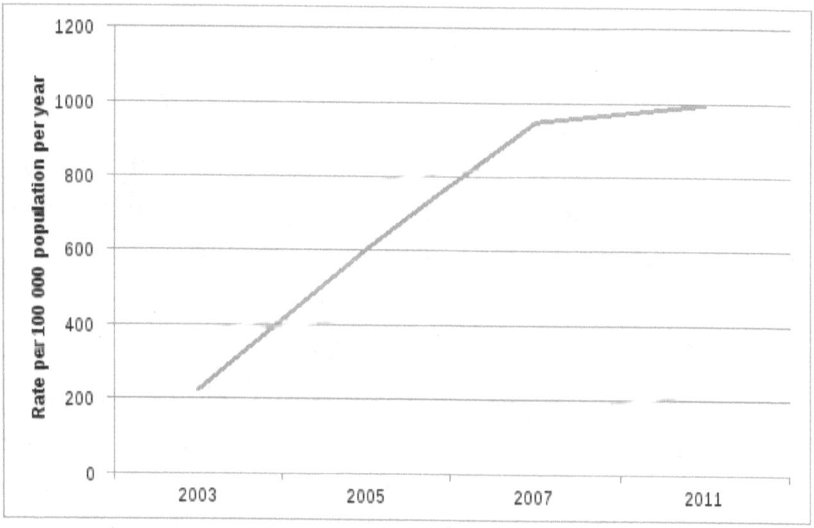

Interestingly, self-reported proportion of adults who reported having had tuberculosis in South Africa seems to be decreasing, while mycobacterium tuberculosis registration data shows that mycobacterium tuberculosis is increasing (Department of Health, Medical Research Council & ORC Macro, 2004). The self-reported decrease in prevalence of mycobacterium tuberculosis may have been due to the fact that the data in the South African Demographic Health

Survey was based on a smaller sampling frame and a smaller sample size of only 10 080 households, and that the respondents may have been untrue about their mycobacterium tuberculosis history. This data was collected by researchers from the tuberculosis research department of the South African Medical Association, and the inclusion criteria was that women should be between 15 and 49 years of age and men should be between 15 and 59 years of age, and the exclusion criteria was that non- permanent residents of the targeted households did not qualify (Department of Health, Medical Research Council & ORC Macro, 2004). Such a small sample size cannot be generalized to the entire South African population. Moreover, the sample was not a true reflection of the complex demographics of South Africa.

Globally, 3.7% (2.1-5.2%) of new cases were estimated to have MDR-TB in the year 2011, and in South Africa 1.8% (1.4-2.3%) of new cases were estimated to have MDR-TB (World Health Organization, 2012). Since the p value was less than 0.001 ($p<0.001$) for all developing countries and was 0.1 ($p=0.1$) for all developed countries, it implies that the proportion of 1.8% cases of MDR-TB in South Africa was significant, and there was 95 per cent confidence that the proportion fell between 1.4 per cent and 2.3 per cent.

The number of deaths due to mycobacterium tuberculosis among individuals who are HIV positive continues to increase, and in 2011 alone, there was an estimated 430 000 deaths globally (World Health Organization, 2012). Au-Yeung et al (2011) had conducted a cross-national systematic assessment to compare the mortality rates in HIV infected individuals and HIV/TB co-infected individuals in the years 2006 to 2008, and the Pearson coefficient was 0.352, the Chi-squared value was 4.05 with a significant level $p = 0.04$. This implies that there was a positive linear relationship between mortality of HIV infected individuals and HIV/TB co-infected individuals. Since $p<0.05$, it implies that there was a significant correlation between

HIV mortality and HIV/TB co-infection mortality. Since the Pearson coefficient = 0.352, it means that the coefficient of determination r^2 = 0.124, which shows that only 12 per cent of the HIV mortality is explained by HIV/TB mortality, and the remaining 88 per cent of the variation is left unexplained.

Strengths of secondary data sources:

There are a number of strengths of the secondary data sources that one has used. Firstly, the fact that the data is already there or is available saves a lot of time than when primary data is collected (Bruce et al, 2008). There is no need to spend time planning for the research design or physically conducting data collection or following-on subjects as the data is already there ready to be consumed.

Secondary data is usually less expensive compared to primary data collection (Cohen et al, 2007). By collecting data from secondary sources cuts costs of the logistics and resources needed when collecting primary data.

Secondary data can eliminate the likelihood of bias because there is no need deal with large populations or sampling frames or the determination of a sample size or complex analyses as the data is already there (Sorensen et al, 1998). In that sense secondary data can reduce the chance of bias. According to George & Landermann (1984), secondary data enables an increased ability to apply multivariate statistical techniques, and greater control and flexibility for the investigator.

Secondary data may be the only available piece of information about a specific issue such as multi-drug resistant tuberculosis. This could therefore be used as a baseline for future primary research for other settings or countries that have not had research conducted about the particular disease such as multi-drug resistant tuberculosis. It is very likely that secondary data provided by an external source such as the World Health Organization will be objective and less biased as it focuses on a global perspective.

Internal secondary data such as from the South African Department of Health and Medical Research has strength because it uses specific breakdowns preferred by the setting, and it is also easy to get more information from the people who conducted the research because they are within the country.

Limitations of secondary data sources:

The fact that the secondary data is already there, could pose problems because the secondary user of the data has no control of the quality of the data, and it may also be difficult to validate such data (Sorensen et al (1998). The fact that the accuracy of the secondary data is dependent on the person who recorded it could be a limitation because if there are mistakes, the secondary user may not be able to manipulate that data or make corrections. Moreover, the secondary data may have been collected by other people for different purposes which may lack specificity and not be compatible with what one would like to use the secondary data for.

There is a possibility that there could be missing data and this could create bias in the analysis (Cohen et al, 2007). The problem here is that it may be difficult to factor for the missing data, as we may not know the real reasons for the missing data as well as the confounding factors related to such secondary data.

It may be difficult to generalize secondary data because we may not be clear about the inclusion and exclusion criteria used (Sorensen et al (1998). Even if the inclusion and exclusion criteria were known, the problem is that such criteria may not be applicable across the board due to inconsistencies of recording such data in different settings or countries. It is possible that secondary data may only show the internal characteristics of a setting or country, and such internal characteristics may not necessarily be extrapolated to describe the entire world as a whole (Sorensen et al (1998). Such variations may occur between provinces within a country, between countries or between regions or continents.

Conclusion

The magnitude of the problem posed by mycobacterium tuberculosis is so enormous that it requires a concerted effort from all countries. Unfortunately, the present scenario indicates that not many countries have strengthened their health systems, and as a result many countries have been found wanting due to lack of effective controls against the spread of mycobacterium tuberculosis. There are strengths and limitations of secondary data sources such as the Global Tuberculosis Report. The main strengths of secondary sources of data are that they are quicker to use, they are cost effective, they are objective, they are less prone to bias, and the fact that the data is complete and available. The main limitations of secondary sources of data are that the user has no control over it which makes it difficult to ascertain its quality, validity and applicability.

REFERENCES

Atun, R., Weil, D.E.C., Eang, M.T. & Mwakyusa, D. (2010) 'Health-system strengthening and tuberculosis control', *The Lancet*, 375 (9732), pp. 2169-2178.

Au-Yeung, C., Kanters, S., Ding, E., Glaziou, P., Anema, A., Cooper, C.L., Montaner, J.S.G., Hogg, R.S. & Mills, E.J. (2011) 'Tuberculosis mortality in HIV-infected individuals: a cross-sectional systematic assessment', *Clinical Epidemiology*, 3, pp. 21-29.

Bruce, N., Pope, D. & Stanistreet, D. (2008) *Quantitative research methods for health research: a practical guide to epidemiology*. Chichester: John Wiley & Sons, Ltd.

Bruce, N., Pope, D. & Stanistreet, D. (2008) 'Routine data sources and descriptive epidemiology'. In: *Quantitative research methods for health research: a practical guide to epidemiology*. Chichester: John Wiley & Sons, Ltd.

Christensen, L.B. (2007) *Experimental methodology*. Boston: Pearson International.

Cohen, J., Bilsen, J., Missinesi, G., Lofmark, R., Addington-Hall, J., Kaasa, S., Norup, M.,Van der Wal, G. & Deliens, L. (2007) 'Using death certificate data to study place of death in 9 European countries: opportunities and weaknesses', BioMed Central Public Health, 7, pp. 283-292.

De Haan, M. (2009) *The health of Southern Africa*. Cape Town: Juta Academic.

Den Boon, S., van Lill, S.W.P., Borgdorff, M.W., Enarson, D.A., Verver, S., Bateman, E.D., Irusen, E., Lombard, C.J., White, N.W., de Villiers, C. & Beyers, N. (2007) 'High prevalence of tuberculosis in previously treated patients, Cape Town, South Africa', *Emerging Infectious Diseases*, 13 (8), pp. 1189-1194.

Department of Health, Medical Research Council & ORC Macro (2004) *South African demographic and health survey, 2003*. Pretoria: Department of Health.

Dheda, K., Shean, K., Zumla, A., Badri, M., Streicher, E.M., Page-Shipp, L., Willcox, P., John, A., Reubenson, G., Govindasamy, D., Wong, M., Padanilam, X., Dziwiecki, A., van Helden, P., Aiwendu, S. Jarand, J., Menezes, C., Burns, A. & Victor, T. (2010) 'Early treatment outcomes and HIV status of patients with extensively drug-resistant tuberculosis in South Africa: a retrospective cohort study', *The Lancet*, 375 (9728), pp. 1798-1807.

Dikeledi, ., Soogreem, T., Kirsten, Z., Bello, B. & Dayal, P. (2009) 'Detection of environmental mycobacterium tuberculosis using rapid and sensitive conventional and real time polymerase chain reaction', *Occupational Health Southern Africa*, 15 (5), pp. 19-24.

Geffken, D.F. , Perry, M.J. & Callas, P.W. (2000) 'Association of occupation and breast cancer mortality in the state of Vermont, 1989-1993', *Massachusetts Journal of Medicine*, 5, pp. 75-79.

George, L.K. & Landermann, R. (1984) 'Health and subjective well-being: a replicated secondary data analysis', *International Journal of Aging & Human Development*, 19 (2), pp. 133-156.

Harling, G., Ehrlich, R. & Myer, L. (2007) 'The social epidemiology of tuberculosis in South Africa: a multilevel analysis', *Social Science & Medicine*, 66, pp. 492-505.

Hattingh, S. & Acutt, J. (2003) *Occupational health management & practice for health practitioners*. Lansdowne: Juta & Co. Ltd.

Hoel, D.G., Ron, E., Carter, R. & Mabuchi, K. (1993) 'Influence of death certificate errors on cancer mortality trends', *Journal of the National Cancer Institute*, 85 (13), pp. 1063-1068.

Joubert, G. & Ehrlich, R. (2007) *Epidemiology: a research manual for South Africa*. 2nd ed. Cape Town: Oxford University Press Southern Africa.

LaDou, J. (2007) *Current occupational & environmental medicine*. New York: McGraw Hill.

Logan, P.A., Coupland, C.A.C., Gladman, J.R.F., Sahota, O., Stoner-Hobbs, V., Robertson, K., Tomlinson, V., Ward, M., Sach, T. & Avery, A.J. (2010) 'Community falls prevention for people who call an emergency ambulance after a fall: randomized controlled trial', *British Medical Journal*, 340, pp. 1-7.

Lonnroth, K., Castro, K.G., Chakaya, J.M., Chauhan, L.S., Floyd, K., Glaziou, P. & Raviglione, M.C. (2010) 'Tuberculosis control and elimination 2010-50: cure, care and social development', *The Lancet*, 375 (9728), pp. 1814-1829.

Nogueira, C. (2009) 'Action on silica, silicosis and tuberculosis – WAHSA experience in Southern Africa', *Occupational Health Southern Africa*, 15 (special issue), pp. 27-35.

Nzimande, P. N. (1996) *Communicable diseases in the African continent*. Pinetown: Alberts Publishers.

Peto, R., Gray, R., Collins, R., Wheatley, K., Hennekens, C., Jamrozik, K., Warlow, C., Hafner, B., Thompson, E., Norton, S., Gilliland, J. & Doll, R. (1988) 'Randomised trial of prophylactic daily aspirin in British male doctors', *British Medical Journal*, 296, pp. 313-316.

Roesel, D. (2006) *Global TB incidence*. Available from: http://ethomed.org/clinical/tuberculosis/firland/latent-tb-faqs

Sorensen, H.T., Sabroe, S. & Olsen, J. (1996) 'A framework for evaluation of secondary data sources for epidemiological research', *International Journal of Epidemiology*, 25 (2), pp. 435-442.

World Health Organization (2012) *Global Tuberculosis Report 2012*. Geneva.

Zumla, A., Kim, P., Maeurer, M. & Schito, M. (2013) 'Zero deaths from tuberculosis: progress, reality, and hope', *The Lancet*, 13 (14), pp. 285-287.

Did you love *Ethics, Qualitative And Quantitative Methods In Public Health Research*? Then you should read *Medicine and Sociology of Health*[1] by Mbuso Mabuza!

[2]

This book aims to critically examine the contribution of the discipline of sociology and a number of theoretical sociological perspectives, from classical to postmodern to our understanding of health, illness and medicine. Specifically, the broad areas covered in this book include: critical perspectives on key sociological theories and themes in the sociology of health and illness including the profession of medicine, doctor-patient relationships, medicalisation, challenges to biomedicine, lay perspectives and beliefs, contemporary aspects such as health promotion, the new genetics, and complementary and alternative medicine. It is my hope that this book will enable you the

1. https://books2read.com/u/bMx1eV

2. https://books2read.com/u/bMx1eV

reader to reflect critically on the sociological perspectives and concepts to your own context and experience of health, medicine, illness and public health practice. The intriguing discussions are organised in the following chapters: introduction - the history and philosophy of medicine; key concepts in sociology - an introduction; sociological theory - from classical to modern; sociological theory - critical and contemporary perspective; doctors, nurses and patients - working for health; challenging the ;medical model' - lay perspectives and the contribution of sociology; health promotion and sociology; complementary and alternative medicine (CAM); Sociology and the "New Genetics".

Also by Mbuso Mabuza

A Healthy Mind And Best You: Achieving Great Results in Every Aspect of Your Life
Purposeful And Better You
Sustainable Development Calls for Effective Strategic Leadership for Efficient Health Systems
Health Promotion In Low Socioeconomic Settings
Medicine and Sociology of Health
Qualitative Methods In Public Health Research
Epidemiological Research
Ethics, Qualitative And Quantitative Methods In Public Health Research
Global Health Disaster Management
Global Health Policy And Programme Challenges
The Journey of Life Has a Gift of Purpose

About the Author

Dr Mbuso Mbuza is a multi-skilled professional who has more than ten years of experience leading demanding health and business portfolios in different countries. Mbuso's expertise is in bio-chemistry, medicine, international public health, occupational health, organisational health and wellbeing, integrative and holistic healthcare, health systems strengthening and sustainability, epidemiology and disease prevention, project management, strategic leadership, research and development, innovations, monitoring and evaluation, impact assessments, editing, academic writing, creative writing, reviewing, publishing, art and design, and entrepreneurship, among others.

www.ingramcontent.com/pod-product-compliance
Lightning Source LLC
LaVergne TN
LVHW090354210926
848480LV00002B/339